THE PRODUCTION AND

APPLICATION OF NEW

INDUSTRIAL TECHNOLOGY

Related works by Edwin Mansfield

THE ECONOMICS OF TECHNOLOGICAL CHANGE

INDUSTRIAL RESEARCH AND TECHNOLOGICAL INNOVATION

RESEARCH AND INNOVATION IN THE MODERN CORPORATION

EDWIN MANSFIELD

JOHN RAPOPORT

ANTHONY ROMEO

EDMOND VILLANI

SAMUEL WAGNER

FRANK HUSIC

The Production and Application of New Industrial Technology

W · W · NORTON & COMPANY · INC · NEW YORK

FIRST EDITION

Library of Congress Cataloging in Publication Data
Main entry under title:
The Production and application of new industrial technology.
Includes bibliographical references and index.
1. Technological innovations—United States—
Addresses, essays, lectures. 2. Research, Industrial—
United States—Addresses, essays, lectures.
I. Mansfield, Edwin.
HC110.T4P76 338'.06'0973 77-9281

ISBN 0-393-09168-6

Published simultaneously in Canada by George J. McLeod Limited, Toronto
Printed in the United States of America

1 2 3 4 5 6 7 8 9 0

Contents

Preface

This book is the latest in a series of volumes summarizing the results of a continuing set of studies of the economics of technological change that my co-workers and I have been carrying out. Previous books in this series are *The Economics of Technological Change* (Norton, 1968), *Industrial Research and Technological Innovation* (Norton, for the Cowles Foundation for Research in Economics at Yale University, 1968), and *Research and Innovation in the Modern Corporation* (Norton, 1971). The present book extends and builds on the results of these previous works, and presents, as well, findings regarding areas untouched by the earlier books. In particular, it contains new findings concerning the social and private rates of return from industrial innovations, the nature of the development process, the determinants of development costs, the relationship between innovation, development, and size of firm, the organizational and strategic factors associated with probabilities of success in industrial research and development, and the determinants of the speed of application of new technology.

I am extremely grateful to the National Science Foundation, which supported these studies, and to the Center for Advanced Study in the Behavioral Sciences at Stanford, where, as a Fellow in 1971–1972, I worked on some aspects of them. I am also indebted to the many firms who took the time and trouble to provide us with the data and expertise without which these projects could not have been carried out. For example, the large industrial firm discussed in Chapters 8 and 9 provided us with an extraordinarily rich body of data, and commented in detail on our findings; and the major R and D executives of several chemical companies went over Chapters 3 to 5. In all, our empirical results are based on data obtained from literally hundreds of firms and research organizations.

Preliminary versions of material contained in this book were presented in papers I gave at the American Economic Association, the International Economic Association meetings in San Anton and Moscow, the U.S.-U.S.S.R. Symposium on the Economics of Information, the Conference on Research in Income and Wealth, the Econometric Society, the National Science Foundation, the Eastern Economic Association, and in testimony before Congressional committees. I am indebted to various discussants of these papers for their comments.

In addition, I am grateful to William Comanor of the University of Western Ontario, Henry Grabowski of Duke University, Zvi Griliches of Harvard University, and Richard Nelson of Yale University for their comments on the manuscript. Finally, I want to thank the *Review of Economics and Statistics,* the *Quarterly Journal of Economics,* the *Journal of Business, Management Science,* and the International Economic Association* for allowing us to reproduce material that first appeared in their pages.

Frank Husic, Anthony Romeo, and Edmond Villani were graduate students at the University of Pennsylvania. In the case of Romeo and Villani, their work on these studies constituted their doctoral dissertations, which I had the pleasure of directing. George Beardsley is a co-author of Chapters 8 and 9; his contribution stems too from his doctoral dissertation which was part of this project. Chapter 2 is the work of Samuel Wagner and myself. Chapter 3 is largely the work of Frank Husic, although Peter Simon was responsible for checking and, in some cases, extending Husic's data. (Simon's thesis, which is referred to in Chapter 6, was also carried out as part of this project.) The first half of Chapter 4 was done by John Rapoport and myself, while the second half was done by Villani, who was also responsible for Chapter 5. Chapter 7 was written by Romeo. Chapters 8 and 9 were done by Rapoport, Romeo, Wagner, George Beardsley, and myself, and I did Chapters 1, 6, and 10 as well as the conversion and adaptation of the parts into a book. The responsibility for the topics we explored and the approaches we used is, of course, mine.

Edwin Mansfield

Philadelphia
January 1977

*Part of Chapter 6 was published in my paper in B. Williams, ed., *Science and Technology in Economic Growth,* London: Macmillan, 1973, which was the printed record of the International Economic Association's San Anton conference.

THE PRODUCTION AND APPLICATION OF NEW INDUSTRIAL TECHNOLOGY

1 THE PRODUCTION OF NEW INDUSTRIAL TECHNOLOGY

1. Technological Change and the Economy

Technological change is undoubtedly one of the most fundamental and important forces affecting the American economy. New technology—consisting of advances in knowledge concerning the industrial and agricultural arts—influences the rate of growth of productivity and living standards, the rate at which environmental pollution grows or subsides, the ease with which we can cope with our energy problems, the competitiveness of American products in world markets, the availability and nature of jobs in many industries, the balance of military power, the nature and structure of particular industries, and the fortunes of individual firms. In recent years, there has been considerable interest among economists in learning more about the process of technological change and its effects on the economy. This upsurge in interest is a response both to the importance of the subject and to the fact that, until recently, the study of technological change was neglected by economists.

In this opening chapter, we provide a brief description of some of the salient features of the process of technological change in American industry. The concepts, facts, and findings presented in this chapter are needed to understand the studies described in the rest of the book. We begin by defining and characterizing industrial research and development (R and D). Then we describe the organization and management of industrial R and D, as well as its risks and estimated profitability. We turn next to the project-selection decision and the size of cost and time overruns, after which we describe the innovation process and time-cost trade offs. Finally, we take up the effects of market structure on the rate of innovation, the nature and importance of the diffusion process, and the current status of technological forecasting.[1]

1. Some sections of this chapter are based on or closely parallel parts of E. Mansfield, *The Economics of Technological Change* (New York: Norton, 1968); E. Mansfield, J.

2. *Industrial Research and Development*

The modern economy is characterized by large investments in research and development. To take a recent year, in 1976, research and development carried out by American firms amounted to about $26 billion. Industrial R and D includes activities of three broad types: basic research, applied research, and development. Basic research is devoted entirely to the creation of new knowledge for its own sake. For example, the mathematician who tries to prove a certain theorem, without any particular application in mind, is performing basic research. Firms carry out some basic research, but, as would be expected, it accounts for a relatively small proportion of their R and D work. In contrast to basic research, applied research is expected to have a specific practical payoff. For example, projects might be carried out to determine whether lasers might be useful for a certain purpose or whether a new type of material might be superior to an old material in a particular product. The distinction between applied and basic research thus comes down to the aim of the work rather than the nature of the activities involved, and is frequently not very precise, although it is widely used.

Development attempts to reduce research findings to practice. Relatively minor development projects try to bring about slight modifications of existing products or processes. Major development projects try to bring into being entirely new types of products and processes. By the time a project reaches the development stage, much of the uncertainty concerning its technical feasibility has been eliminated, but there often is considerable uncertainty concerning development cost, development time, and the profitability of the result. The tasks that are carried out in development depend, of course, on the nature and purpose of the development project. Frequently, prototypes must be designed and constructed, or pilot plants must be built. Relatively little detailed research has been carried out to describe and analyze the anatomy of the development process in particular industries. One purpose of this book is to present the results of some research of this sort pertaining to polymers.

The distinction between research and development often is not clear-cut. The distinction between them relates primarily to the orientation of the work, the degree and type of uncertainty, and the length of time work can be expected to go on without any practical payoff. Research is oriented toward the pursuit of new knowledge, whereas development is oriented toward the capacity to produce a

---

Rapoport, J. Schnee, S. Wagner, and M. Hamburger, *Research and Innovation in the Modern Corporation* (New York: Norton, 1971); E. Mansfield, "Technological Forecasting," in T. S. Khachaturov, ed., *Methods of Long-Term Planning and Forecasting* (London: Macmillan, 1976); and E. Mansfield, "The Economics of Industrial Innovation: Major Questions, State of the Art, and Needed Research," in P. Kelly and M. Kranzberg, *Technological Innovation: A Critical Review of Current Knowledge,* forthcoming.

particular product. Research generally entails more uncertainty of outcome, and requires more time for completion, than development. Although research and development are not the same, they do tend to shade into one another.

At the outset, several facts concerning America's investment in research and development should be noted. First, from the end of World War II until the late 1960s, the nation's expenditures on research and development increased at a relatively rapid rate. But from 1968 to 1975, when inflation is taken roughly into account, evidence presented by the National Science Foundation seems to indicate a decline in the total expenditures on R and D in constant dollars. The number of scientists and engineers engaged in R and D also declined—from 560,000 in 1969 to 520,000 in 1973. Some policy makers and economists have been concerned that this decline may have a detrimental effect on the nation's future rate of economic growth.

Second, much of the research and development performed by industry is financed by the federal government. For example, in 1974, the government financed about 40 percent of the R and D performed by industry. Much of the federally funded R and D performed by industry is related to national defense and space; but in recent years there has been some shift in emphasis toward civilian areas like energy and the environment. Industry-financed research and development, which totaled about $14 billion in 1974, is concentrated heavily in five industries—electrical equipment and communication, chemicals and allied products, machinery, motor vehicles, and aircraft and missiles. These five industries accounted for about 80 percent of all industry-performed research and development in 1973.[2]

Third, most of the money spent on R and D goes for development, not research. Only about one third of the nation's R and D expenditures go for research, and only about one eighth go for basic research. Development accounts for a particularly large proportion of the total in the machinery, electrical-equipment, and aircraft and missiles industries. In most industries, the principal objective of most R and D expenditures is to improve existing products, rather than to develop entirely new products or processes. For example, a survey carried out in 1975 found that about 36 percent of the responding firms' R and D expenditures were aimed at developing new products, 50 percent were aimed at improving existing products, and 14 percent were aimed at developing new processes.[3] (However, lest there be misunderstanding, it is important to recognize that one industry's products may be part of another industry's processes.)

2. National Science Foundation, *Science Indicators 1974* (Washington, D.C.: Government Printing Office, 1975); and *Science Resources Studies Highlights,* National Science Foundation, May 21, 1976.

3. *Business Plans for R and D Expenditures, 1975–1978,* (New York: McGraw-Hill, 1975).

3. The Organization of Industrial Research and Development

Firms organize their R and D activities in quite different ways. In large, decentralized firms, there is frequently a central research laboratory, which emphasizes research, and separate R and D departments in each of the operating divisions, these departments doing product and process development of a more immediate nature. There are advantages in separating the more fundamental work from the more short-term, since the organization, style of leadership, and constraints that are appropriate for more short-term work are often inappropriate for more fundamental work. However, it obviously is important that this separation does not result in too little communication between the advanced researchers and the people who are wrestling with problems of current design, manufacturing, and marketing. Otherwise, research findings may not be properly utilized, and research projects may not be chosen with proper reference to the firm's commercial needs.

According to a survey of 19 laboratories in the chemical, drug, petroleum, and electronics industries, about 58 percent of these laboratories have separate research functions and development functions, each with its own staff and facilities. A relevant variable in explaining why some laboratories tend to separate research from development whereas others do not is the relative equality of research expenditure and development expenditure. As research expenditure and development expenditure become more nearly equal, the need for the separation of research from development intensifies, while if research expenditure or development expenditure is very small relative to the other, there is much less need for separation.[4]

Within a firm's central research laboratory, there is often a grouping of scientific personnel by academic disciplines or subjects. For example, the survey of chemical, drug, electronics, and petroleum firms indicated that research was organized more often along disciplinary than product or functional lines. The subject areas, which often parallel the fields of learning at educational institutions, naturally depend on the problems predominant in the firm. This disciplinary kind of organization makes it relatively easy for the researcher to keep in touch with others in the same subject field, it encourages him to keep up to date in his science, and it makes it easier to recruit outstanding scientists to the company.

With regard to projects at the development end of the R and D spectrum, scientists and engineers are generally organized on a product basis, each group being composed of various disciplines needed to solve problems in the development, design, or manufacture of a particular set of products. For example, the survey cited above indicated that development was organized more often along

4. Mansfield *et al.*, *Research and Innovation*, *op. cit.*

product than disciplinary or functional lines. In large, decentralized firms, the product-specialized R and D groups may be physically decentralized, each being attached to the relevant operating division. Typically, the divisional R and D manager reports to the general manager of his division, who has day-to-day control over the division's R and D staff and activities. The corporate R and D director reviews and coordinates the R and D programs of all the operating divisions, looking particularly at the technical quality and promise of the work. Organization of applied work on a product basis has a number of advantages, the most important being that it directs the work toward the realities of the market place and shortens the communication links between the people in R and D and the design groups who can use their findings in the divisions.

In handling multidisciplinary problems, special project teams are often formed. For example, members of various departments of the central research laboratory may be combined with a development group for a limited period of time to accomplish a particular task. Organization by project results in numerous benefits. Since the objective is clear and the researchers know that they will be evaluated in terms of their success in achieving it, the motivation to adhere to relevant work is strengthened. Coordination of the work is easier than when parts of the work are being done in a number of separate and separated units. Moreover, the interaction of several disciplines and the periodic shifting of personnel may help to promote productivity.[5]

4. The Management of Research and Development

The administration and control of research and development tends to be looser and more informal than in most other areas of industry. Although there may exist a rather formidable formal control structure, many of the decisions tend in fact to be made quite informally. In research, but not development, it is generally believed that administrators should resist the temptation to manage projects in much detail. Usually, management's job is limited to providing general guidelines and goals, selecting capable persons, and providing a climate conducive to good work. But this does not mean that management does not—and should not—guide research. Whereas overmanagement of research is likely to lead to technical hack work, totally unguided research is likely to lead to waste and frustration. Firms that do well in this area learn to guide research in the directions most likely to contribute to company goals, yet not constrain it so that it produces excessively narrow and short-range work.

Each company must decide, in the light of expected scientific and commer-

5. See A. Stanley and K. White, *Organizing the R. and D Function* (New York: American Management Association, 1965); and R. Anthony, *Management Controls in Industrial Research Organization* (Cambridge, Mass.: Harvard University Press, 1952). Also, see chap. 3 in Kelly and Kranzberg, *op. cit.*

cial developments, the scientific areas where it will concentrate its efforts, those where it will merely keep in touch with the scientific community, and those where it will virtually ignore developing knowledge. Top management's ultimate control over the scope, emphasis, and efficiency of research lies largely in the selection of key personnel. Research and general managers decide in what broad fields to do research, and staff the program accordingly. Within these fields, they tend to rely heavily on the judgment of researchers to select specific inquiries and particular approaches to the problem.

In development, management plays a more dominant role. In part, this is because there is less uncertainty. Management can define the problems and goals more closely, and it can be more confident of a correlation between resources utilized and achievement. Another important reason for the greater role of management is the high cost of development. If a project proceeds beyond the exploratory stages, its costs begin to increase considerably. Moving from a design study to the later stages of development work typically involves greater labor requirements for working out and testing the details of the design. Testing the full system involves considerable expense, particularly when the production of test items requires many special tools and buildings. From a sample of large projects in widely different product fields, it appears that the cumulative number of man-years devoted to a project can be approximated reasonably well by a logistic function of time.[6] Due to their more expensive nature, development projects are scrutinized carefully, and major commitments are not made until the design concept seems both technically feasible and economically attractive.

The management of research and development is presently far from a science. Numerous propositions are put forth, but few are testable and practically none has been tested in any formal way. One purpose of this book is to present the results of some exploratory econometric studies bearing on this important topic. The following seven points summarize a few of the propositions in the literature, and help to illustrate the sorts of managerial problems that exist in this area. First, it is frequently stated that a company's research aims should be clearly spelled out and communicated to the scientists, and that its research should be undertaken in broad areas relevant to its economic goals. If a firm simply hires teams of scientists and allows them to do research in their favorite fields, the results may be novel and interesting, but they are unlikely to have much commercial value to the company.

Second, it is often pointed out that top management's responsibilities do not end with the definition of the objectives of research and development. Management must also make sure that its organization performs certain critical planning functions. Research and development should be responsive to technological

6. P. Norden, "Curve Fitting for a Model of Applied Research and Development Scheduling," *IBM Journal of Research and Development* (July 1958).

flows from three sources: the company's competitors, the company's present and prospective customers, and the scientific community in general. Despite the difficulties in technological forecasting described in section 11, the firm must attempt to forecast, as best it can, the new developments that will come from these sources.

Third, management must pay proper attention to the transfer of new technology into operations. This is sometimes an extremely difficult task, as we shall see in Chapter 2. On the one hand, this transfer may be restricted because the researcher does not recognize the commercial implications of his work or does not want to release his findings until he has covered more of its scientific ramifications. On the other hand, operating groups sometimes resist new technology on the ground that it is "impractical" or because, although it is desirable in the long run, it will add to their short-run marketing or engineering costs. Various devices, such as the movement of people across the interface between R and D and other departments and the creation of new product committees and departments, have been used to help reduce the resistance to the transfer of new technology into operations. More will be said on this score in later chapters.

Fourth, particularly in the case of large military and space projects, it is often claimed that managers tend to underestimate the uncertainties involved, that they do not maintain enough flexibility, and that they settle too quickly on a particular design. When there is considerable uncertainty, it often pays to proceed sequentially, maintaining flexibility and using the experience that is gained to reduce uncertainty. This is illustrated by the use of parallel R and D efforts, discussed in section 5.

Fifth, it is important to prevent the isolation of research from development. In a company where this is the case, the research people may be divorced from practical applications and their work may have little practical value. On the other hand, according to many observers, there are dangers if no line is drawn between research and development. The program may become dominated by short-term, product-oriented projects, and the exploration of fundamentally new areas may be neglected.

Sixth, it is frequently pointed out that a reasonable limit should be placed on the amount of "fire-fighting" activities to be done by researchers. In most companies, management tends to value technical service to customers and to the company's manufacturing divisions as one of the most important products of its technical groups. Sometimes management feels that whenever an emergency service call comes in, longer-range R and D projects can be dropped temporarily. Too much work of this sort produces resentment among scientists and engineers, and detracts considerably from a laboratory's power to produce important new ideas in the long run.

Seventh, it is also important to recognize that the firm's R and D department is only one of many sources of new ideas. There is sometimes a tendency for R and D managers to focus too much attention on new developments arising from their laboratories, and to neglect developments in other firms and industries. If something is "not invented here," it may be neglected or resisted.[7]

5. *Returns and Risks*

For a variety of reasons, it is very difficult to make accurate forecasts of the returns from a particular R and D project. Thus, firms find it difficult to determine with precision how much they ought to spend on research and development. In the short run, they sometimes maintain a fairly constant ratio between R and D expenditures and sales. In the longer run, they adjust their desired ratio of R and D expenditures to sales in response to changes in the perceived profitability of research and development, changes in the profitability of the firm, and changes in the amount spent by their competitors.

Some tentative and experimental estimates have been made of the marginal rate of return from R and D expenditures in a variety of firms and industries. Griliches, Mansfield, Minasian, and Terleckyj have carried out studies of this sort based on the idea that R and D expenditures, as well as labor and capital, should be included as inputs in the production function. In general, the results of these studies suggest that the marginal rate of return from R and D expenditures has been relatively high. However, for reasons discussed by all of these authors, these estimates should be viewed with great caution.[8]

7. See Stanley and White, *op. cit.;* S. Kingsburg *et al.*, "Organizing for Research," in C. Heyel, ed., *Handbook of Industrial Research Management* (New York: Reinhold, 1959); S. Marcson, *The Scientist in American Industry* (New York: Harper & Row, 1960); J. Quinn, "Long-Range Planning of Industrial Research," *Harvard Business Review* (July 1961); J. Quinn and R. Cavanaugh, "Fundamental Research Can Be Planned" *Harvard Business Review* (January 1964); H. Shepard, "Patterns of Organization for Applied Research and Development," *Journal of Business* (January 1956); N. Kaplan, "Some Organizational Factors Affecting Creativity," *IRE Transactions on Engineering Management* (1960) D. Marquis, "Organization and Management of R and D," in *Proceedings of the 17th Conference on Administration of Research* (Denver: University of Denver, 1964); T. Allen and S. Cohen, "Information Flow in an R and D Laboratory," *Administrative Science Quarterly* (March 1969); N. Baker and J. Freeland, "Structuring Information Flow to Enhance Innovation," *Management Science* (September 1972); and various issues of the *IEEE Transactions on Engineering Management* and *R and D Management*.

8. Z. Griliches, "Research Expenditures, Education, and the Aggregate Agricultural Production Function," *American Economic Review* (December 1964); E. Mansfield, *Industrial Research and Technological Innovation* (New York: Norton, for the Cowles Foundation for Research in Economics at Yale University, 1968); J. Minasian, "Research and Development, Production Functions, and Rates of Return," *American Economic Review* (May 1969); N. Terleckyj, *Effects of R and D on the Productivity Growth of*

One major purpose of this book is to present the results of some further studies of the private and social returns from investments in innovation. These results are contained in Chapters 8 and 9. In contrast with the highly aggregated and rather indirect analyses cited in the previous paragraph, we shall look in detail at the returns from a sample of specific industrial innovations. Moreover, estimates will be made of the gap between social and private rates of return, a topic that (as we shall see) may be of relevance to the formulation of public policy toward civilian technology.

It is also worth noting that studies have been made of the relationship between a firm's R and D expenditures and its output of significant inventions, these studies being limited to the major firms in the chemical, petroleum, and steel industries. When the size of the firm is held constant, the number of significant inventions carried out seems to be highly correlated with the size of its R and D expenditures. Thus, although the output from an individual R and D project is obviously very uncertain, there appears to be a close relationship in the long run between the amount the firm spends on R and D and the total number of important inventions it produces.[9]

One of the most important characteristics of research and development is that it is relatively risky. For example, an intensive study of the R and D programs of three firms (one in chemicals, two in proprietary drugs) showed that, on the average, the probability that an R and D project would result in an economically successful product or process was only about 0.12.[10] In other words, about one out of eight projects turned out to be economically successful. To help cope with such uncertainty, some R and D projects use parallel approaches, that is, they carry forward several alternative designs or approaches to the problem. In cases where it is difficult to tell at the outset which approach is best, and where it is relatively inexpensive to try out additional approaches, this sort of strategy is quite appropriate.

However, it is important to note that, to a very great extent, the risks in industrial research and development are commercial, not technical. The technical risks involved in the bulk of industrial research and development—outside military and other government-financed areas—are often quite modest. For example, among nineteen laboratories in the petroleum, chemical, electronics, and drug industries, the average probability of technical completion for a project was greater than fifty-fifty. To a large extent, this seems to be due to the fact that the bulk of the R and D projects are aimed at fairly modest advances in the state of the art. Particularly in laboratories that devote a large percentage of their re-

---

Industries: An Exploratory Study (Washington, D.C.: National Planning Association, 1974); and Z. Griliches, "Returns to Research and Development in the Private Sector," Conference on Research in Income and Wealth, 1975.

9. Mansfield, *Industrial Research, ibid.*

10. Mansfield *et al.*, *Research and Innovation, op. cit.*

sources to development rather than research, the technical risks seem relatively small.

On the other hand, the commercial risks—the risk that the new or improved product or process resulting from a technically successful R and D project will not seem sufficiently profitable to warrant its being commercially introduced or applied, or that, if it is introduced or applied, it will not be an economic success—often are substantial. For example, in the three firms cited above, about 40 percent of the R and D projects that were begun were not technically completed. Of those projects that were technically completed, 45 percent were not commercialized, presumably because commercialization did not seem profitable. Of those projects that were commercialized, about 60 percent did not earn an economic profit. Thus, it is much more likely that a project will achieve its technical aims (this probability being about 0.60 in these firms) than that, if it achieves these aims, it will earn an economic profit (this probability being about 0.20 in these firms). In other words, it is much more likely that the laboratory can solve the technical problems involved in developing the new or improved product or process than that it will be economically worthwhile to solve these problems—in the sense that the new or improved product or process will turn out to be a commercial success.[11]

Finally, it is important to note that research and development is aimed at reducing uncertainty. That is, R and D is aimed at learning. At the outset of a project, there generally is uncertainty concerning the feasibility and cost of developing a product or process with certain desired characteristics. And this uncertainty is due in part to imperfect knowledge regarding various technical and engineering relationships. During the course of the R and D effort, many of the basic uncertainties tend to be resolved, at least in part. Indeed, the basic purpose of many of the activities carried out in research and development is to resolve uncertainty—at a reasonable cost.

6. *Project-Selection Decisions*

Project selection is obviously an important aspect of the decision-making process in industrial research and development. The way in which such decisions are made will vary, depending on whether it is a research or development project. As a project moves from research toward the market, it is subjected to more intensive scrutiny from both a technical and an economic angle. When projects are in the early research phase, the screening of proposals may be relatively quick and informal, since the costs at this stage are relatively small and it is difficult to predict the outcome of such work. However, as projects enter develop-

11. *Ibid.*

ment, where costs and predictability are greater, a much more detailed economic evaluation of proposals is generally carried out.

Quantitative techniques—generally adaptations of well-known capital-budgeting schemes—have been proposed by many authors to help industrial laboratories in project selection. According to various surveys, although many of the larger laboratories have tried them, the more complicated of these techniques have not been accepted widely. One important problem in using these techniques is that cost overruns and time overruns seem to be large in civilian R and D, as well as in military programs. For example, in a proprietary drug firm, the average ratio of actual to expected development cost was 2.11, and the average ratio of actual to expected development time was 2.95. A comparison of the overruns in drug firms with the overruns in a dozen major military projects indicates that the cost overruns tend to be larger in military work than in the drug industry, while the time overruns tend to be smaller in military work than in the drug industry. The factors that seem to influence a project's cost overrun are much the same in the drug industry as in military work.

One purpose of this book is to present the results of an econometric study of the determinants of development costs in polymers. This study should help to indicate the extent to which a formal econometric model, based on a highly disaggregated view of the development process, may be able to forecast the development cost of a particular project. The results should be of interest, since it clearly is important for firms and government agencies to obtain a better capacity to forecast the costs of developing a new product or process.

In making decisions concerning project selection, it is important for a firm to look outward for ideas and to avoid focusing too much attention exclusively on its own inventions. As noted in section 4, there is sometimes a tendency for R and D managers to neglect developments that are "not invented here." This is unfortunate because external sources of technology are frequently very important. For example, about one half of the major new products in the ethical drug industry in 1935–1962 were based on discoveries made outside the innovating firm. And about one third of the innovations included in a sample of chemical, electronics, and machinery innovations were based on specific technology derived from outside the innovating firm. More will be said on this score in subsequent chapters.[12]

12. For further discussion of project-selection techniques, see M. Cetron, J. Martino, and L. Roepcke, "The Selection of R and D Program Content: Survey of Quantitative Methods," *IEEE Transactions on Engineering Management* (March 1967); W. Souder, "Utility and Perceived Acceptability of R and D Project Selection Models," *Management Science* (August 1973); R. Seiler, *Improving the Effectiveness of Research and Development* (New York: McGraw-Hill, 1965); the papers by Robert E. Gee, Richard E. Faust, Eric S. Whitman, Edward F. Landau, and Richard L. Meek in *Research Management,* September 1971; Mansfield, "Technological Forecasting," *op. cit.;* and the works cited there.

7. *The Innovation Process*

Economists define an innovation as the first commercial application of a new or improved process or product. There are many models or descriptions of the phases involved in the process leading up to an innovation. For example, Myers and Marquis[13] say that the process begins with a new idea that involves the recognition of both technical feasibility and potential usefulness or demand. But this new idea, or design concept, is only the formulation of a problem that seems worth pursuing. The next step, according to Myers and Marquis, is the problem-solving activity—the search processes and research and development—that, if successful, results in a solution to the technical and commercial problems related to the idea. Finally, there is the stage when the new device is introduced commercially. Of course, there is no guarantee that a given project will go through all of these stages. As we have seen, most projects do not clear all of the technical and commercial hurdles required for completion and implementation. Moreover, as Langrish *et al.*[14] and other economists point out, the sequence of activities leading up to an innovation often fails to conform to any simple linear model.

Many studies have attempted to shed light on the extent to which firms are induced to undertake innovations by technical opportunities or by market factors. In other words, to what extent are innovations triggered by market needs, and to what extent are they triggered by technological opportunities? Obviously, this is not an easy distinction to make, and data based on it are not easy to interpret. But based on the studies conducted by Carter and Williams, Goldhar, Langrish *et al.*, Myers and Marquis, the Materials Advisory Board of the National Academy of Sciences, Baker *et al.*, and other investigators,[15] a much larger proportion of innovations are initially stimulated by the recognition of market and/or production need than by recognition of a technological opportunity. Specifically, about three quarters of the innovations included in these studies seemed to be stimulated initially by a recognition of such need. Schmookler's findings, based on patent statistics, point in essentially the same direction.[16]

13. S. Myers and D. Marquis, *Successful Industrial Innovations* (Washington, D.C.: National Science Foundation, 1969). Also, see OECD, *The Conditions for Success in Technological Innovation* (Paris, 1971).

14. J. Langrish *et al.*, *Wealth from Knowledge* (New York: Macmillan, 1972).

15. C. Carter and B. Williams, *Industry and Technical Progress* (New York: Oxford, 1957); J. Goldhar, "An Exploratory Study of Technological Innovation" (unpublished, 1970); Materials Advisory Board, *Report of the Ad Hoc Committee on Principles of Research-Engineering Interaction* (Washington, D.C.: National Academy of Sciences, 1966); N. Baker *et al.*, "The Effects of Perceived Needs and Means on the Generation of Ideas for Industrial R and D Projects," *IEEE Transactions on Engineering Management* (1967).

16. J. Schmookler, *Invention and Economic Growth* (Cambridge, Mass.: Harvard, 1966).

Studies have also been carried out to determine the sources of the major information inputs that evoked the design concept for the innovation, as well as the information used in the subsequent problem-solving activity. According to Myers and Marquis, in about 25 percent of the cases they studied, an information input initiated active work on the problem. With regard to the solution of problems already being worked on by the firm, the major information inputs which contributed to the solution of the problem were predominantly general in nature, widely diffused, and generally accessible. (Active search, analysis, experimentation, and outside help were involved in getting the major information input in only about 24 percent of the cases.) Personal experience and personal contacts were the principal sources of information for the cases they studied.

Despite the advances that have been made in recent years, there are many interesting aspects of the innovation process that have received far too little attention. One purpose of this book is to present the results of a study of the factors associated with the percentage of innovation costs that go for R and D. For a variety of reasons (explained in Chapter 4), this is an interesting topic. Also, we provide very detailed data concerning the sources of process and product innovations in the chemical industry since 1930. Based on these data, we can study some questions that are of considerable significance from the point of view of public policy. Since innovation is a key stage in the process leading to the full evaluation and utilization of new technology, it is important that the innovation process be better understood.

8. Time-Cost Trade Offs

Before leaving the innovation process, it is worth pointing out that considerable research has been carried out in recent years concerning time-cost trade offs in the innovation process. A product often can be developed and brought to market more quickly if more money is spent during the course of the innovation process. Estimates of the time-cost trade-off function have been made for twenty-nine innovations in chemicals, electronics, and machinery on the basis of a series of interviews with the managers who had the principal responsibility for the innovations. The data suggest that a downward-sloping convex function will usually be a reasonable approximation to the time-cost trade-off function (in the relevant range). The empirical support for the hypothesis that the slope is negative is very strong. There is somewhat more question about the hypothesis of convexity, but in this sample at least, a convex shape is strongly suggested in a large proportion of the cases.

The elasticity of cost with respect to time is the expected percentage increase in cost due to an expected 1-percent reduction in the duration of the project. The available evidence indicates that this elasticity becomes greater as the duration of the project is pushed closer and closer to its minimum value. When the duration of the project is less than 30 percent above its minimum value, this elasticity averages about 1.6 in this sample of innovations. When the duration of the project is about 30 to 80 percent above its minimum value, this elasticity averages about 0.5. According to these data, a project's elasticity of cost with respect to time depends on the extent to which the project advances the state of the art and on the size of the firm carrying out the project: Projects that attempt greater advances in the state of the art and that are carried out by larger firms tend to be more costly than others to speed up.

The evidence also suggests that overlap—the beginning of one stage of the innovation process before a previous stage is completed—may be a major reason for the increased expected costs associated with a decrease in expected time. In this sample of innovations, the amount of overlap tended to increase as an innovation's time to completion was pushed closer and closer to the minimum time to completion. Moreover, there was a tendency for bigger firms to overlap stages to a greater extent than smaller ones. In general, it was felt by the managers that were interviewed that increased overlap between two stages would tend to increase the cost of the later stage. The most commonly mentioned reason for this was characterized by one manager as the increased likelihood of "engineering-change notices." The problem arises when work in the later stage is critically dependent on information or results obtained in the earlier stage. When there is considerable overlap, the later stage must be started before the results of the work in the earlier stage are known. Estimates or tentative results must be used as the basis for the later work. If these estimates prove to be incorrect, the expense of correcting mistakes and redoing work may be considerable.[17]

9. *Technological Change and Market Structure*

A number of important questions pertain to the relationship between an industry's market structure—the number and size distribution of firms—and its rate of technological change: Is it true, as some claim, that innovation is now the province of very big firms? Is considerable monopoly power, as measured by high market concentration,[18] conducive to a rapid rate of technological change? In recent years, economists have begun to study these questions in a systematic way. Although we are far from having complete or definitive answers, quite a few generalizations seem warranted, based on the evidence gathered to date.

The first point that must be made is that the role of the small firm is important at the stage of invention and the initial, relatively inexpensive stages of R and D.

17. For more extensive discussions of these results, see Mansfield *et al., Research and Innovation, op. cit.*

18. A commonly used measure of concentration is the percent of the market accounted for by the largest four firms.

Based on the findings of Jewkes *et al.*, Scherer, Hamberg, Mueller, and other economists,[19] it is clear that small firms and independent inventors play a large—perhaps a disproportionately large—role in conceiving important new ideas and major inventions. Moreover, although the development of these ideas often requires more resources and different kinds of management skills than some of these small firms have, the investment required for development and innovation is seldom so great or so risky that only the largest firms in the industry can do the developing or the innovating. Indeed, our own studies of the drug, petroleum, steel, and coal industries show that, in all of these industries, the firms that carried out the largest number of innovations—relative to their size—were not the biggest firms. (In the steel and drug industries, they tended to be surprisingly small firms.)[20]

Based on studies by Scherer, Mansfield, and others,[21] it is clear as well that the biggest firms generally do not spend more on R and D, relative to their size, than somewhat smaller firms. There seems to be a threshold effect. A firm has to be a certain size to spend much on R and D as defined by the National Science Foundation. But beyond a certain point, increases in size no longer bring any disproportionate rise in R and D expenditures. This seems to be true in all industries studied, with the exception of chemicals. Of course, the threshold varies from industry to industry, but, according to Scherer, increases in size beyond an employment level of about 5,000 employees is generally not accompanied by a more than proportional increase in innovation inputs or outputs. Thus, the evidence does not seem to indicate that giant firms devote more resources, relative

19. J. Jewkes *et al.*, *The Sources of Invention*, 2nd ed. (New York: Norton, 1969); D. Hamberg, "Invention in the Industrial Laboratory," *Journal of Political Economy* (April 1963); F. M. Scherer, *Industrial Market Structure and Economic Performance* (Chicago: Rand McNally, 1971); and W. Mueller, "The Origins of the Basic Inventions Underlying DuPont's Major Product and Process Innovations, 1920–1950," in *The Rate and Direction of Inventive Activity* (New York: National Bureau of Economic Research, 1962). Also see E. Mansfield, "Federal Support of R and D Activities in the Private Sector," in *Priorities and Efficiency in Federal Research and Development* (Joint Economic Committee of Congress, 1976).

20. However, our results concerning the drug industry are based on data ending in the 1960s. More recent data suggest that the largest drug firms account for a substantially larger share of the innovations than before. In part, this seems to be attributable to the increase in development costs due to the 1962 amendments to the Food, Drug, and Cosmetic Act. See H. Grabowski, *Drug Regulation and Innovation* (Washington, D.C.: American Enterprise Institute, 1976).

21. Mansfield, *The Economics of Technological Change, op. cit.;* Scherer, *op. cit.;* H. Grabowski, "The Determinants of Industrial Research and Development," *Journal of Political Economy* (March 1968); and works cited there. For a recent review article, see M. Kamien and N. Schwartz, "Market Structure and Innovation: A Survey," *Journal of Economic Literature* (March 1975).

to their size, to inventive and innovative activites than their somewhat smaller competitors. Moreover, there is some evidence that, per dollar of R and D, the biggest firms get less inventive and innovative output than smaller firms.

Of course, firm size and market structure are two different things. A big firm need not be a monopolist or anything close to it. From the point of view of antitrust policy, it is particularly important to explore the relationship between the extent of concentration in an industry and the industry's rate of technological change. Far too little work has been carried out in this area to support firm or confident conclusions. But certain patterns are beginning to emerge. In general, most studies of the relationship between industrial concentration and the rate of technological change seem to conclude that a slight amount of concentration may promote more rapid invention and innovation. Very fragmented, splintered industries like construction do not seem to be able to promote a rapid rate of technological advance. But beyond a moderate amount of concentration, further increases in concentration do not seem to be associated with more rapid rates of technological advance. In part, this may be due to less competitive pressure and fewer independent loci for decision making.[22]

Finally, three additional points should be noted. First, there is a great deal of evidence that new firms and firms entering new markets play a very important role in the promotion of technological change. Existing firms can be surprisingly blind to the potentialities of new ideas. They can become bureaucratized and tradition-bound. An important way that their mistakes and inertia are overcome in our economy is through the entry of new firms. Examples of new firms that pioneered the introduction of new products or processes are Polaroid, with self-developing photography, Houdry, with catalytic cracking, and Marconi, with the radio. Moreover, there have been many cases where the threat of entry by new firms or firms in other industries has stimulated existing firms to carry out significant technological advances.

Second, it is generally agreed by economists that the ideal market structure from the point of view of stimulating technological change is one where there is a mixture of firm sizes. Complementarities and interdependencies exist among larger and smaller firms. There is often a division of labor, smaller firms focusing on areas requiring sophistication and flexibility and catering to specialized needs, bigger firms focusing on areas requiring larger production, marketing, or technological resources. There seem to be considerable advantages in a diversity of firm sizes, no single firm size being optimal in this respect. Moreover, the optimal average size is likely to be directly related to the costliness and scope of the inventions that arise.

Third, cases sometimes arise where industries are composed of such small firms, or markets are so fragmented, that technological change is hampered. Firms cannot afford to do much R and D, and they may find it difficult to use new

22. For example, see Scherer, *op. cit.*

techniques. The risks involved in innovation are too great for any single firm. In such cases, it may be good public policy to supplement the R and D supported by the private sector. Of course, a good example of this is in agriculture, where the government has supported a great deal of R and D for many years.

One purpose of this book is to add to the present fund of knowledge concerning the relationship between size of firm and innovation (and development) in an important science-based industry, chemicals. Despite the importance of the chemical industry as a generator of technological progress, relatively little research has been carried out to determine the sources of various major chemical innovations, and the extent to which the largest chemical firms have done a disproportionately large share of the innovating. Another relevant aim of this book is to investigate the effect of industrial concentration on an innovation's rate of diffusion. More will be said on this score in the following section.

10. The Diffusion Process

The diffusion process—the process by which the use of an innovation spreads—is of great social importance. For example, in the case of a process innovation, the rate of diffusion determines how rapidly productivity increases in response to the new process. In a free-enterprise economy, firms and consumers are free to use new technology as slowly or as rapidly as they please, subject, of course, to all the constraints imposed by the market place. Diffusion, like the earlier stages in the creation and assimilation of new methods and products, is essentially a learning process. However, instead of being confined to a research laboratory or to a few firms, the learning takes place among a considerable number of users and producers.

When the innovation first appears, potential users are uncertain of its nature and effectiveness, and they tend to view its purchase as an experiment. Sometimes considerable additional research and development is required before the innovation is successful; sometimes, despite attempts at redesign and improvement, the innovation never is a success. Information regarding the existence, characteristics, and availability of the innovation is disseminated by the producers through advertisements and salesmen; information regarding the reaction of users to the innovation tends to be disseminated informally and through the trade press. Learning takes place among the producers of the innovation, as well as the users. Early versions of an innovation often have serious technological problems, and it takes time to work out these bugs. During the early stages of the diffusion process, the improvements in the new process or product may be almost as important as the new idea itself. Moreover, when a new product's design is stabilized, costs of production generally fall in accord with the so-called "learning curve." That is, unit costs of production decrease as the producers gain experience and learn by doing. For example, in the case of aircraft production during World War II, there is some evidence that average labor requirements tended to be reduced by about 20 percent for each doubling of cumulated output.[23]

In recent years, a number of studies have been carried out concerning the diffusion process. Based on this work, we are able to describe the rate at which various innovations have spread, the kinds of firms that have been relatively quick, or relatively slow, to begin using them, and the rate at which individual firms have substituted various new techniques for older ones. Moreover, models have been constructed and tested to help explain and forecast an innovation's rate of diffusion. A major aim of this book is to extend and further test these models. Besides examining the effect of industrial concentration on the rate of diffusion, we shall look at how the amount spent on R and D by the potential adopters seems to affect the rate of diffusion. Further, we shall devote some attention to the determinants of the rate at which a product (as distinct from a process) innovation is imitated. Despite their importance, none of these topics has received much attention in the literature.

11. Technological Forecasting

A final subject that should be touched on is technological forecasting. It is generally agreed that technology is a difficult variable to forecast because there is so much uncertainty concerning what will be produced by R and D efforts and concerning what breakthroughs will occur, and when. Nonetheless, technological forecasting has been attempted by many firms and government agencies. The techniques most frequently used by technological forecasters seem to be simple intuitive projections, the Delphi method, and trend extrapolation. These techniques seem crude, even by the standards of the social and management sciences, and it seems unlikely that the results can be at all accurate.

To improve the state of the art, some experimentation is going on with more sophisticated models. For example, input-output analysis has been used to forecast future input requirements; econometric models of aggregate output, which include R and D expenditures, have been used for certain kinds of technological forecasting; and econometric models of the diffusion process are beginning to be used for forecasting purposes by government agencies and business firms. One purpose of this book is to describe a particular case where an econometric diffusion model was used for this purpose.

The fact that existing technological forecasting techniques are crude does not mean that firms and government agencies do not have to make such forecasts.

23. W. Hirsch, "Firm Progress Ratios," *Econometrica* (April 1956); and K. Arrow, "The Economic Implications of Learning By Doing," *Review of Economic Studies,* June 1962.

Just as there is no way to avoid forecasting the economic future—explicitly or implicitly—so there is no way to avoid forecasting the technological future. But this does not mean that it is necessarily worthwhile for a firm or government agency to support any formal work in technological forecasting. Whether or not it is worthwhile depends on whether—under the particular set of conditions confronting the firm or agency—the potential gains seem to outweigh the costs. And given the lack of reliable data regarding the likely gains from various kinds of technological forecasting, this is not an easy comparison to make.

If done properly, technological forecasting involves much more than just science and engineering. One cannot estimate the probability that a particular technology will come into being on the basis of technological considerations alone. Economic, social, and political considerations often play an equally important role. Moreover, one cannot decide how an organization's technological resources should be allocated and utilized on the basis of technological considerations alone. Clearly, economic, social, political, and other considerations are involved here as well. Thus, regardless of what sort of mechanism a firm or agency uses to supply its technological forecasts—whether that mechanism be a formal group or an informal one, and whether it use intuitive or more objective techniques—it is important that the mechanism be properly tuned in to the economic, social, and political environment, as well as to the relevant science and technology.

12. Summary and Conclusions

New technology, consisting of advances in knowledge concerning the industrial and agricultural arts, has a fundamental and important influence on the economy. To produce various kinds of new technology, firms in the United States carry out about $26 billion worth of research and development, about 40 percent of which is financed by the federal government. The bulk of the company-performed R and D is concentrated in five industries—electrical equipment and communication, chemicals and allied products, machinery, motor vehicles, and aircraft and missiles. Most of the work is development, not research. In large, decentralized firms, there is frequently a central research laboratory, which emphasizes more fundamental and long-term work, and separate R and D departments in each of the operating divisions, these departments doing product and process development of a more immediate nature.

The administration and control of research and development tend to be looser and more informal than in most other areas of industry. However, because there is less uncertainty in development than in research, management plays a more dominant role in development. The management of research and development is presently far from a science. Numerous propositions are put forth, but few are testable and practically none has been tested in any systematic way. One central, and difficult, problem in this area is to estimate accurately the returns from a firm's research and development, and to forecast the prospective profitability of proposed projects.

Research and development is relatively risky, but it is important to recognize that, to a considerable extent, the risks are commercial, not technical. Since the bulk of the industrial R and D projects are aimed at fairly modest advances in the state of the art, the technical risks often are quite modest. Substantial overruns occur with respect to cost and time in civilian as well as military work, with the result that quantitative project-selection techniques often are difficult to apply.

An innovation is defined as the first commercial application of new technology. A much larger proportion of industrial innovations seems to be stimulated initially by the recognition of market and/or production need than by recognition of a technological opportunity. In general, there seems to be a trade off between time and cost in developing and bringing a new product to market. The nature of this trade off varies with the characteristics of the innovation and of the innovator. The innovator plays an important social role. Based on studies of the steel, petroleum, coal, and drug industries, the firms that carried out the largest number of innovations, relative to their size, were not the biggest firms. Moreover, the biggest firms generally do not spend more on R and D, relative to their size, than somewhat smaller firms.

The diffusion process is of great social significance, since an innovation cannot have its full effects until it is widely used. Like the earlier stages of the process by which new technology is produced and applied, the diffusion process is a learning process. In recent years, there has been considerable attention paid to technological forecasting. Existing techniques in this area seem rather crude, and experiments are currently being carried out to find more satisfactory techniques. To develop better methods, it seems likely that researchers and practitioners must improve their basic understanding of the process of technological change.

2 ORGANIZATIONAL AND STRATEGIC FACTORS ASSOCIATED WITH PROBABILITIES OF SUCCESS IN INDUSTRIAL R AND D

1. Introduction

This chapter presents the results of some econometric and statistical studies carried out to test a number of hypotheses concerning the effects of various organizational and strategic factors on a firm's probabilities of success in its R and D program. The results are based on data obtained from twenty major firms in the chemical, drug, petroleum, and electronics industries. We begin by defining three probabilities of success. Then we study the effects of several organizational and strategic factors on each of these probabilities of success, this study being based on data from sixteen of the firms. Next, we look more intensively at the effect of two factors: the degree of integration between R and D and marketing, and the degree of formality and quantification of the R and D project-selection system. This study is based on very detailed data from three of the firms. Finally, we investigate the extent to which there may be an underutilization of R and D results due to inadequate understanding and receptivity in other parts of the firm. This study is based on data from eighteen of the firms. So far as we can tell, this seems to be the first systematic study of interfirm differences in such probabilities of success.

2. *Probabilities of Success: Definition and Measurement*

In this chapter, we are concerned entirely with three probabilities: the probability of technical completion, the probability of commercialization (given technical completion), and the probability of economic success (given commercialization). The *probability of technical completion* is the probability that an R and D project will achieve its technical objectives. The *probability of commercialization* (*given technical completion*) is the probability that a technically complete R and D project will be commercialized—that is, that there will be a full-scale marketing or application of the new or improved product or process beyond a test-market or pilot-plant trial. The *probability of economic success* (*given commercialization*) is the probability that a commercialized R and D project will yield a rate of return (on the R and D costs plus any additional investment made to introduce the innovation) in excess of what was available from other (non–R and D) investment alternatives. Note that the product of these probabilities equals the probability that an R and D project begun by the firm will be an economic success. Also, note that basic research is not included here (for obvious reasons) and that the bulk of the projects are development projects.

Probabilities of this sort are commonly used to measure the riskiness of a firm's R and D program, as well as how well it manages and utilizes its R and D resources. Clearly, these probabilities are by no means ideal measures: indeed, they suffer from a number of well-known difficulties.[1] But in a field where

1. These probabilities indicate the percentage of projects—or percentage of R and D dollars, since, as we shall see below, the projects are weighted by their cost—that went for completed or successful projects. Although they are of interest, these probabilities may be misleading if firms with low values of these probabilities tend to obtain a much higher average rate of return from economically successful projects than firms (in the same industry) with high values of these probabilities. Existing evidence is too weak to tell us definitely whether or not this is the case, but based on the estimates of the average rate of return from economically successful projects obtained from the firms, there is no evidence that this was true among the firms in our sample.

To see how these probabilities are related to a firm's return from its R and D investment, let N be the number of R and D projects begun by a firm in a particular period, N_t be the number that are technically completed, N_c be the number commercialized, and N_s be the number of economic successes. If R_B is the average cost of an R and D project that is begun but not technically completed, R_t is the average cost of an R and D project that is technically completed but not commercialized, and R_c is the average cost (R and D and

measurement is so difficult, it seems worthwhile to try to determine how much light interfirm differences in these probabilities can shed on differences among firms in the nature of their R and D portfolios and in the quality of their R and D managements. Based on intensive interviews with twenty major firms, we obtained estimates of each of these probabilities for most of the firms. These figures seem reasonably accurate, since they were calculated on the basis of a careful review by each firm of its past performance. In calculating these probabilities, projects were weighted by their cost, since otherwise the results might be unduly influenced by very small projects.

Table 2.1 shows the frequency distribution of each of these probabilities during 1968–1971 for sixteen firms in our sample.[2] (The other firms did not provide

---

other) of a project that is commercialized, then the total investment in R and D and commercialization of R and D results is:

$$I = NR_B + N_t(R_t - R_B) + N_c(R_c - R_t),$$

and the total annual return from this investment equals

$$\pi = N_s\pi_s,$$

where π_s is the average annual profit per economic success. (We assume for simplicity that the average return from economic failures is zero.) Letting P_t equal the probability of technical completion, P_c equal the probability of commercialization (given technical completion), and P_s equal the probability of economic success (given commercialization), it is clear that $P_t = n_t/N$, $P_c = N_c/N_t$, and $P_s = N_s/N_c$. Consequently,

$$I = N[R_B + P_t(R_t - R_B) + P_tP_c(R_c - R_t)]$$

and

$$\pi = NP_tP_cP_s\pi_s.$$

Thus

$$\pi = \frac{I\ P_tP_cP_s\pi_s}{R_B + P_t(R_t - R_B) + P_tP_c(R_c - R_t)}$$

Holding constant I, π_s, R_t, R_B, and R_c, it is obvious that π increases with P_t, P_c, and P_s.

Note, however, that the probabilities we use are generally weighted by project cost. (See note 2). Thus, under the conditions described above, $\pi^* = I^*P_t^*P_c^*P_s^*\pi_s^*$, where π^* is the average annual return from the investment in R and D, P_t^* is the proportion of R and D expenditures spent on technically completed projects, P_c^* is the proportion of the amount spent on technically completed projects that is spent on commercialized projects, P_s^* is the proportion of the amount spent on commercialized projects that is spent on economically successful projects, I^* is the total investment in R and D, and π_s^* is the average rate of return from a dollar spent on an economically successful project. Obviously, if I^* and π_s^* are held constant, π^* increases with P_t^*, P_c^*, and P_s^*.

2. Note several things concerning the sample. (1) Many of these "firms" are major divisions of a large firm, not entire firms. There are advantages, of course, in having the data pertain to less heterogeneous groupings. (2) In the regressions in section 4, not all of the twenty firms could be included because not all provided data on each of the relevant independent and dependent variables. Those that could be included in equation (2.4) are

TABLE 2.1

Estimated Probability of Technical Completion, Commercialization (Given Technical Completion), and Economic Success (Given Commercialization), 1968–1971

PROBABILITY	PROBABILITY OF TECHNICAL COMPLETION	COMMERCIALIZATION (GIVEN TECHNICAL COMPLETION)	ECONOMIC SUCCESS (GIVEN COMMERCIALIZATION)
		(number of firms)	
.90 and over	2	3	3
.80–.89	4	1	1
.70–.79	3	4	2
.60–.69	1	3	4
.50–.59	0	3	1
.40–.49	1	0	0
.30–.49	1	0	0
Less than .30	4	2	0
Total	*16*	*16*	*11**
Average probability	0.57	0.65	0.74

* Estimates could be obtained for only eleven firms.

all the data we requested.) These data show that there is great variability among firms in these probabilities. In some firms, the probability of technical completion is as high as 0.94; in other firms, it is as low as 0.05. In some firms, the probability of commercialization (given technical completion) approaches 1.00; in other firms, it is as low as 0.12. In some firms, the probability of economic success (given commercialization) approaches 1.00; in other firms, it is as low as

included in column 1 of Table 2.1, those that could be included in equation (2.5) are included in column 2 of Table 2.1, and those that could be included in equation (2.6) are included in column 3 of Table 2.1. (3) Tables 2.2, 2.3, and 2.4 are based on those firms that could be included in equation (2.4).

The average probability of technical completion is very similar to that which we found in earlier studies of nineteen industrial laboratories, the latter being 0.57. Also, the average probability of commercialization (given technical completion) is quite similar to that for three laboratories we studied before, the latter being 0.55. However, the probability of economic success is higher than for the three firms included in a previous study, the latter being 0.38. See E. Mansfield, J. Rapoport, J. Schnee, S. Wagner, and M. Hamburger, *Research and Innovation in the Modern Corporation* (New York: Norton, 1971).

Note too that projects are weighted here, but not in Table 2.5, by the size of their R and D expenditures. The results are essentially the same if unweighted data are used instead.

0.50. Even in the same industry, each of these probabilities can be much higher in one firm than in another. These interfirm differences are of interest and importance. An adequate understanding of the factors responsible for these differences would almost surely promote a better understanding of interfirm differences in the nature, riskiness, and productivity of R and D.

3. A Simple Model

To help explain these interfirm differences, we construct a simple econometric model, which hypothesizes that these three probabilities are influenced by the following three variables: First, we would expect that all three of these probabilities would be affected by how quickly R and D projects are evaluated from the point of view of potential market and profit. Some firms allow R and D projects to proceed much further than do other firms before the potential profitability of the project is studied. Table 2.2 shows that, on the average, the firms in our sample permitted about $40,000 to be spent on an R and D project before such a study was made. But there was a great deal of interfirm variability in this respect. Some firms allowed $200,000 to be spent before such a study, whereas other firms spent little or nothing before it.[3]

TABLE 2.2

Amount of Money that Typically Can Be Spent on an R and D Project before Studies Are Made of Market and Profit Potential, Sixteen Firms, 1970

AMOUNT OF MONEY (THOUSANDS OF DOLLARS)	NUMBER OF FIRMS
Less than 10	7
10–24	3
25–49	1
50–99	2
100–149	1
150–199	1
200 and over	1
Total	*16*

3. By a "study" or "evaluation" of the potential market and profitability of the R and D project, we mean a serious attempt by competent professionals to carry out such a task. Such a study need not be very elaborate or expensive, but it must be made by people who are reasonably disinterested and competent, and it must be more than a token effort.

Among the firms in Table 2.2, there is relatively little correlation between a firm's size

In general, there are many arguments for integrating technological considerations with economic considerations relatively early in the game. Unfortunately, one suspects that some firms do not integrate these factors early enough, the result being that some projects with very little potential economic payoff are started and continued too long. And because this is the case, the probability of technical completion is lowered, since more projects are started which are stopped short of technical completion because of poor profit prospects.[4] Also, the probability of commercialization (given technical completion) is lowered because more projects are completed technically before it is recognized that their profit outlook is poor. And the probability of economic success (given commercialization) is lowered, since the firm's portfolio of R and D projects tends to be more poorly geared to economic realities and conditions than would otherwise be the case.

Second, we would expect two of these probabilities to be affected by the percentage of a firm's R and D expenditures that go for "demand-pull" rather than "technology-push" projects. Put somewhat crudely, "demand-pull" projects are designed to satisfy a felt market or management need (or to solve a recognized production problem), whereas "technology-push" projects are designed to find specific uses or markets for promising technology or for a potential product or process. This dichotomy—which, of course, is sometimes extremely blurred in particular cases—is frequently used to characterize R and D projects. Table 2.3 shows that, on the average, about three quarters of the R and D projects carried out by the sixteen firms in our sample were regarded by the firms as primarily "demand-pull," not "technology-push."[5] But there is considerable variation among firms in this percentage. In some firms in the sample, it is as high as 100 percent, whereas in other firms, it is as low as 20 percent.

(as measured by sales) and the amount of money that typically can be spent on an R and D project before studies are made of market and profit potential. Specifically, $\overline{R}^2$ equals 0.11. Moreover, there is even less correlation between the total amount spent annually on R and D by the firm (or relevant part of the firm) and the amount of money that typically can be spent on an R and D project before studies are made of market and profit potential.

4. Of course, it may also be the case that the probability of technical completion will increase because more projects will be carried to technical completion before being stopped because of poor prospects. But it seems unlikely that this effect will outweigh that cited in the text, except in unusual cases.

5. This percentage is almost precisely the same as the average percentage obtained by Carter and Williams, Goldhar, Langrish *et al.*, and Myers and Marquis. See K. Pavitt, *The Conditions for Success in Technological Innovation* (Paris: OECD, 1971). Also, see N. Baker *et al.*, "The Effects of Perceived Needs and Means on the Generation of Ideas for Industrial R and D Projects," *IEEE Transactions on Engineering Management* (December 1967). In interpreting numbers of this sort, it is important to bear in mind that this dichotomy is more blurred than is sometimes acknowledged.

TABLE 2.3

Percentage of R and D Expenditures Devoted to "Demand-Pull" Projects, Sixteen Firms, 1968–1971

PERCENTAGE	NUMBER OF FIRMS
90 and over	8
80–89	1
70–79	2
60–69	2
50–59	1
40–49	1
Less than 40	1
Total	*16*

In general, we would expect the probabilities of technical completion and of commercialization to be higher in firms that devote more of their R and D resources to "demand-pull" projects, rather than to "technology-push" projects. The probability of technical completion would be expected to be higher for "demand-pull" projects because it is less difficult and risky to complete a project where the problem or need is reasonably well specified than one where the need or market must be found or created. Similarly, we would expect that the probability of commercialization (given technical completion) would be higher for a "demand-pull" project, since if technical completion is attained, it is much more likely that a market really exists for the innovation, and that other parts of the firm will accept and utilize the R and D results. With regard to the probability of economic success (given commercialization), it is less obvious that this factor will be of importance. Once projects have passed all of the tests required before a firm decides to go ahead with commercialization, it is not obvious that "demand-pull" projects are any better (or worse) bets than "technology-push" projects.[6]

Also, one might expect some of these probabilities to be influenced by the extent to which the firm's R and D portfolio is based on ideas coming from the R and D personnel, as distinct from the marketing personnel and other parts of the firm (or its suppliers and customers). Of course, it is not easy—or sometimes even possible—to identify exactly where particular ideas have come from; and in some cases, there is no single source. Table 2.4 shows the estimates made by our sample of sixteen firms. On the average, about two thirds of the R and D projects

6. However, to the extent that "technology-push" projects tend to be more ambitious and aimed at larger potential payoffs, the probability of economic success (given commercialization) may decrease as the percentage of R and D expenditures devoted to "demand-pull" projects increases. But the data provide no evidence of this.

TABLE 2.4

Percentage of R and D Projects Where the Basic Idea Stemmed from the R and D Personnel, Sixteen Firms, 1968–1971

PERCENTAGE	NUMBER OF FIRMS
90 and over	3
80–89	1
70–79	5
60–69	4
50–59	0
40–49	1
30–39	1
Less than 30	1
Total	*16*

stemmed from the R and D personnel.[7] However, there is considerable variation among firms in this percentage. In some firms, about 95 percent of the projects stemmed from the R and D personnel, whereas in other firms, about 20 percent of the projects stemmed from this source.

In general, one might expect that firms with a relatively high percentage of R and D projects from the R and D personnel would have a relatively low probability of technical completion. This is because the R and D personnel tend to be attracted by, and interested in, projects that are technically more demanding, more ambitious, and more risky from a technical viewpoint than those suggested by the marketing and other personnel. Also, the projects stemming from the R and D personnel may be less firmly based on market realities than those submitted by the marketing (and other) personnel, so they may be more likely to be stopped short of technical completion because of lack of potential profitability. On the other hand, because the R and D personnel understand the technical aspects of R and D projects so well, they tend to be better judges than other people of what is technically feasible. Thus, one might argue that a high percentage of R and D projects from the R and D personnel might be associated with a relatively high

7. The average percentage in Table 2.4 is remarkably close to the results we obtained in an earlier survey, the average in the earlier study being 62 percent. See Mansfield *et al.*, *Research and Innovation, op. cit.* In part, this may be due to the fact that about half of the firms included in this study were also included in the previous survey.

It may also be worth pointing out that, among the firms in Table 2.4, there is only a low correlation between the percent of projects stemming from the R and D personnel and the percent of R and D expenditures devoted to "demand-pull" projects. However, as would be expected, what correlation there is is negative.

probability of technical completion.[8] It is difficult to judge the relative importance of these conflicting tendencies. We shall test the hypothesis that, on the average, the former factors more than offset the latter, although, on a priori grounds, this is by no means obvious.

One might also expect the probability of economic success (given commercialization) to be higher among firms with a relatively high percentage of R and D projects from the R and D personnel. Basically, this is because the projects stemming from the R and D personnel tend to be more ambitious and more fundamental than those stemming from the marketing and other personnel. If they can be completed technically and if they pass the commercial tests that precede the decision to commercialize, one might expect them to be more likely to pay off handsomely. (Of course, whether the increase in the probability of market success [given commercialization] is big enough to offset the decrease in the probability of technical completion can vary from case to case.) As in the case of the hypothesis presented in the previous paragraph, there are factors running counter to this hypothesis. (For example, one might argue that projects stemming principally from outside R and D tend to be better geared to the realities of the market.) As in the previous paragraph, we feel that this is an interesting hypothesis to test, even though we are by no means committed to its validity.[9]

4. Tests of the Model

Summarizing the hypotheses described in the previous section, we assume that

$$T_i = \alpha_0 - \alpha_1 A_i + \alpha_2 N_i - \alpha_3 R_i + z_i, \tag{2.1}$$

$$C_i = \alpha_4 - \alpha_5 A_i + \alpha_6 N_i + z_i', \tag{2.2}$$

$$M_i = \alpha_7 - \alpha_8 A_i + \alpha_9 R_i + z_i'', \tag{2.3}$$

where T_i is the probability of technical completion for the ith firm, C_i is the probability of commercialization (given technical completion) for the ith firm, M_i is the probability of economic success (given commercialization) for the ith firm, A_i

8. It should be recognized, however, that if the idea for a project originates outside R and D, and if the R and D people feel that it is not technically feasible, the project generally will not be accepted. Thus, the fact that a relatively large percentage of projects stems from outside R and D does not mean that the R and D personnel do not regard them as feasible.

9. Also, the probability of commercialization (given technical completion) may be associated with the percentage of projects stemming from the R and D personnel, but there is no evidence that this is the case.

is the amount (in thousands of dollars) that can be spent on an R and D project in the ith firm before studies are made of market and potential profit, N_i is the percentage of the ith firm's R and D expenditures that go for "demand-pull" rather than "technology-push" projects, R_i is the percentage of the ith firm's R and D projects that stem from the R and D personnel, and z_i, z'_i, and z''_i are random error terms. The intercepts—α_0, α_4, and α_7—may vary, of course, from industry to industry, reflecting the differences among industries in risk and other relevant factors.

To test these hypotheses, we use least-squares to estimate the αs, the results being:

$$T_i = \begin{Bmatrix} 49.7 \\ 23.9 \end{Bmatrix} - \underset{(2.82)}{0.21}\, A_i + \underset{(4.92)}{0.92}\, N_i - \underset{(3.53)}{0.71}\, R_i, \tag{2.4}$$

$$C_i = \begin{Bmatrix} 32.6 \\ 65.6 \end{Bmatrix} - \underset{(2.69)}{0.23}\, A_i + \underset{(2.53)}{0.48}\, N_i, \tag{2.5}$$

$$M_i = 53.2 - \underset{(2.46)}{0.17}\, A_i + \underset{(1.79)}{0.41}\, R_i. \tag{2.6}$$

In equation (2.4), the top figure in brackets pertains to the twelve nondrug firms and the bottom figure pertains to the four drug firms. In equation (2.5), the top figure in brackets pertains to the fourteen nonelectronics firms and the bottom figure pertains to the two electronics firms. Such interindustry differences seem quite sensible in light of previous studies.[10]

The results are in accord with our hypotheses. All of the regression coefficients have the expected signs and are statistically significant. Moreover, these equations can explain a substantial proportion of the variation in the three probabilities, T_i, C_i, and M_i. Specifically, equation (2.4) explains about 80 percent of the variation in T_i, $\bar{R}^2$—corrected for degrees of freedom—being about 0.78. Equation (2.5) explains about one half of the variation in C_i, $\bar{R}^2$—corrected for degrees of freedom—being about 0.54. And equation (2.6) explains about 40

10. In previous studies, we found that the probability of technical completion tends to be lower in the drug industry than in the others. This seems to be due to the nature of pharmaceutical research and the inherent risks (and rewards). See Mansfield *et al.*, *Research and Innovation, op. cit.* As for the relatively high probability of commercialization (given technical completion) in the electronics industry, this may be because fewer major steps following technical completion must be carried out in electronics to commercialize an innovation.

Note that other industry dummy variables were tried and discarded in all three equations—(2.4), (2.5), and (2.6). All of the dummies were statistically nonsignificant except for the one pertaining to drugs in equation (2.4) and the one pertaining to electronics in equation (2.5). Also, note that T_i, C_i, and M_i are percentages, not proportions.

percent of the variation in M_i, $\overline{R}^2$—corrected for degrees of freedom—being about 0.38.

Of course, we would not expect these equations to account for practically all the variation in these probabilities, since many important variables must be omitted because there is no adequate way to measure them. Among these omitted variables is the quality of a firm's scientists and engineers. To a great extent, the effects of such variables are contained in the random error terms, z_i, z'_i, and z''_i. Since our data pertain only to major firms, which are reasonably homogeneous with respect to variables like the kinds of work they are doing and the quality of their personnel, the variation of the random error terms would not be expected to swamp the explained variation, but it would be expected to be big enough to keep $\overline{R}^2$ at moderate levels.

5. *Implications of the Results*

To the extent that firms want to increase these probabilities of success, our findings seem to suggest the desirability of early investigations of the profit potential of R and D projects. According to equations (2.4) to (2.6), one can increase T_i, C_i, or M_i by about 4 percentage points simply by decreasing A_i by \$20,000. Of course, this is too bald an interpretation of the results. To a considerable extent, A_i is an index which is associated with many other characteristics of a firm's organization and behavior. It is a surrogate for how rapidly the firm attempts to gauge the profit potential of an R and D project—or put the other way around, it is a surrogate for how long the firm will let a project go on without looking carefully at its profit potential. Also, our results may be due in part to a correlation between a firm's preferences with regard to risk and its value of A_i.[11]

But recognizing these considerations, it nonetheless seems likely that one major reason for our results is that better-managed firms—which, *all other things equal,* tend to have higher values of these probabilities—generally have relatively small values of A_i. After all, unless marketing and profit considerations are brought into the picture relatively early in the game, it is inevitable that much R and D will be wasted because it will be directed at problems that, although technically interesting, are not commercially worthwhile. Successful innovation entails the fulfillment of both technological and economic objectives. Unless both are considered at a relatively early stage, the chance of success is reduced. Of course, the results of a study of profit potential may be relatively unreliable at an early stage of the project, and we do not mean to imply that the results should be used in a mechanical way or given more weight than they deserve. But used

11. We made some (crude) attempts to test this hypothesis in interviews with the firms, and the results seemed to indicate that it is not true. But nonetheless the possibility should be borne in mind that this hypothesis partially explains the observed effect of A_i.

properly, even rough estimates may help to weed out projects that have practically no chance of commercial success (even though they may be of technical interest) and to result in improved formulation of various projects.

Turning to the observed effects of N_i on these probabilities, it is important to stress that our results do not indicate that a firm should necessarily increase its value of N_i. It is quite likely that "technology-push" projects tend to be more risky than "demand-pull" projects. Thus, firms with less aversion to risk may invest a larger proportion of their R and D expenditures in such projects. Since their probabilities of success are relatively low, we may find a relationship of the sort indicated by equations (2.4) and (2.5). However, if this is the case, this relationship does not indicate necessarily that these firms are inefficient or irrational. It may indicate only that they are willing to take bigger risks than the others. (In addition, it may indicate that the results of "technology-push" projects are harder to sell to marketing, production, and other parts of the firm than those of "demand-pull" projects.)

Our findings also seem to indicate that there is a trade off between the probability of technical completion and the probability of economic success (given commercialization). Holding other factors constant, the probability of technical completion goes down as the firm invests a larger percentage of its R and D expenditures in technically more ambitious and fundamental projects. This seems to be a major reason why, in equation (2.4), the probability of technical completion decreases as more of the R and D portfolio stems from the R and D personnel. On the other hand, the probability of economic success (given commercialization) seems to increase as the firm invests a larger percentage of ts R and D expenditures in technically more ambitious and fundamental projects. This seems to be a major reason why, in equation (2.6), the probability of economic success (given commercialization) increases as more of the R and D portfolio stems from the R and D personnel. It is important that firms recognize that this trade off exists, and that they decide on rational grounds how far they want to go in reducing one probability to increase the other.

6. Integration between Marketing and R and D, Project-Selection Systems, and Probabilities of Success

As a first step toward a more detailed evaluation of the effects of a firm's organization and the nature of its R and D project-selection system on its probabilities of success, we put forth two hypotheses concerning the effects of these factors. First, we would expect that the extent of the communication and cooperation between the marketing and R and D personnel will influence some of the probabilities discussed in previous sections. Judging from the results of past research,[12] successful innovation depends in an important way on R and D being integrated with marketing. The R and D personnel must be able and willing to respond to the marketing personnel's needs, and marketing personnel should be involved in R and D project selection. In contrast, the R and D personnel in some firms have been quite remote at times from the marketing personnel, and have marched to the beat of quite a different drummer, the result being that the R and D output has been more poorly mated with market conditions than would otherwise have been the case.

More specifically, we would expect that closer integration of R and D and marketing would result in a higher probability of commercialization (given technical completion). One reason why projects are technically completed but not commercialized is that the market for the potential new product turns out to be inadequate. Closer integration of marketing and R and D makes it more likely that such projects are stopped prior to technical completion, but the effect on the probability of technical completion is not clear. On the one hand, closer integration may mean that fewer projects with relatively poor profit potential are selected in the first place, with the result that a smaller proportion is terminated for nontechnical reasons prior to technical completion. On the other hand, because fewer projects will be taken all the way to technical completion and then shelved because of poor profit prospects, it may mean that this proportion will increase. Similarly, the effect of closer integration on the probability of economic success (given commercialization) is not clear. It may mean that marketing will be better able to exploit new products, thus raising this probability. Or it may result in an R and D portfolio that is composed of a larger proportion of less ambitious projects—projects which, if technically successful, have a relatively small payoff. This may lower this probability.

Second, we would expect that some of these probabilities of success may be influenced by the way in which R and D projects are selected. In recent years, some firms have tried to use formal, quantitative project-selection techniques based on estimated rates of return, pay-out periods, and other such criteria. These techniques may use the results of the studies of market and profit potential discussed in section 3 (assuming that the latter studies are carried out before the relevant projects are begun); but they may not be the same as the latter studies, since they tend to involve much more formal estimation and numerical manipulation, and they tend to have much more ambitious objectives. We would expect

12. See C. Freeman, "A Study of Success and Failure in Industrial Innovation," and E. Mansfield, "Determinants of the Speed of Application of New Technology," both in the International Economic Association's *Science and Technology in Economic Growth* (New York: Macmillan, 1973). Also, see P. Lawrence and J. Lorsch, "New Management Job: The Integrator," *Harvard Business Review* (November 1967); their "Differentiation and Integration in Complex Organizations," *Administrative Science Quarterly* (June 1967), and Mansfield *et al., Research and Innovation, op. cit.*

that the use of such quantitative project-selection techniques may increase the probability of commercialization (given technical completion). Such techniques are designed to promote a more careful and complete evaluation of a project's commercial potential. If they are successful in this regard, the probability of commercialization (given technical completion) should increase, since fewer projects will be technically completed and then shelved because of poor profit potential.

On the other hand, we suspect that the use of such quantitative techniques may result in a decrease in the probability of economic success (given commercialization), because they may tend to push a firm's R and D portfolio in the direction of less ambitious projects with more modest potential payoff. Since it is much easier to estimate the returns from such projects, it is also easier to get them through a screen based on quantitative selection techniques. But if they can be technically completed and commercialized, the more ambitious and fundamental projects with the really big potential payoffs are much more likely to be economic successes. Thus, we would expect that the use of such techniques may reduce the probability of economic success (given commercialization). Finally, the effect of such techniques on the probability of technical completion is not clear. The probability may increase because the use of such techniques may result in a less (technically) risky R and D portfolio. Or a larger proportion of projects that would otherwise be carried to technical completion and then shelved for nontechnical reasons may be stopped prior to technical completion.

7. *Experimental Design and Tests of Hypotheses*

To test these hypotheses, we use data concerning three firms, designated by A, B, and C. These three firms were chosen because they belonged to the same two-digit industry (chemicals and allied products), were of roughly the same size, and spent very close to the same amount annually on R and D. However, during the 1960s, they experienced quite different sorts of major reorganizations. In each firm, a reorganization changed the extent to which R and D and marketing were integrated, and, at the same time, a significant change was made in the formality and degree of quantification of the project-selection system. Figure 2.1 shows the direction of both the reorganization and the change in project-selection system in each firm. As you can see, the reorganizations resulted in a closer integration of marketing and R and D in firms A and C, but in less integration of marketing and R and D in firm B. At the same time, the reorganizations resulted in a more formal and quantitative project-selection system in firms A and B, and a less formal and quantitative project-selection system in firm C. (For details, see note 13.)[13]

13. All three firms are among *Fortune*'s five hundred largest industrial corporations. Two of them were very close in terms of sales; the third was larger. During the relevant

FIGURE **2.1**

Changes in Degree of Integration of R and D and Marketing and in Degree of Formality and Quantification of Project-Selection System, Firms A, B, and C.

	CHANGE IN PROJECT-SELECTION SYSTEM	
Change in Organizational Relationship between R and D and Marketing	Informal and Nonquantitative to Formal and Quantitative	Formal and Quantitative to Informal and Nonquantitative
Closer integration	Firm A	Firm C
Less integration	Firm B	

These three firms provided us with data concerning more than 330 individual R and D projects that occurred from three to seven years before the reorganization to five to eight years after it. In each firm, we computed the probability of technical completion, the probability of commercialization (given technical completion), and the probability of economic success (given commercialization) for the period before the reorganization and for the period after the reorganization. A comparison of each probability's value before the reorganization with its value

---

time period, they spent almost precisely the same amount on R and D. In each of these firms, it was evident that the integration between marketing and R and D changed as indicated in Figure 2.1. In firms A and C, the management felt that marketing had too small a role in the R and D program management, and the R and D department was reorganized to make R and D more responsive to the needs of marketing. Communication channels and networks were established between them, and marketing's input to R and D decision making increased substantially. On the other hand, in firm B, R and D became much more independent from marketing as a consequence of the reorganization. It tended to establish its own criteria and priorities regarding projects without paying nearly as much attention to marketing as before the reorganization.

At the same time, changes took place in the firms' project-selection techniques. Before the reorganizations, no formal project evaluation was undertaken in either firm A or firm B. In each firm, projects apparently were selected on the basis of the intuitive judgment of a selection committee, there being little detailed analysis of expected benefits and costs and only a passing comparison with other projects. After the reorganization, a quantitative project-selection system—based on estimates of project cost, duration, and probability of technical completion and market estimates of sales and product life—was instituted in both firms. In firm C, on the other hand, the project-selection system became less formal and less quantified after the reorganization, partly by neglect and partly by design. A short while after the reorganization, project selection at firm C became highly subjective and informal.

In all three firms, other organizational changes of various types also occurred during the relevant period, but there seemed to be general agreement that these were the most important.

afterward should provide some information concerning the effects of the reorganization. However, such comparisons are complicated by the fact that two changes—a change in the extent of integration of marketing and R and D, and a change in the project-selection system—occurred simultaneously. We would like to extract what information we can concerning the effect of each separate type of change.

To accomplish this, we make paired comparisons between the firms. Firms A and B experienced similar changes in project-selection systems, but opposite changes in the degree of integration between marketing and R and D. Thus, to test whether closer integration resulted in a higher probability of commercialization (given technical completion), we see whether this probability increased more (or decreased less) in firm A (which experienced closer integration) than in firm B (which experienced lesser integration). Any such difference cannot be due to the change in project-selection system, since this change was essentially the same in both firms. Of course, a difference of this sort, if it exists, may be due to other causes than the change in the degree of integration between marketing and R and D, since we cannot hold all other variables constant. But it is worthwhile to determine whether the data are in accord with our hypotheses.

As to the effect of the project-selection system, firms A and C experienced similar changes in the degree of integration between marketing and R and D, but opposite changes in the nature of the project-selection system. Thus, to test whether a more quantitative project-selection system increases the probability of commercialization (given technical completion), we see whether this probability increased more in firm A than in firm C. And to test whether a more quantitative project-selection system decreases the probability of economic success (given commercialization), we see whether this probability increased less (or decreased more) in firm A than in firm C. Any such differences cannot be due to the change in the degree of integration between R and D and marketing, since this change was essentially the same in both firms. Again, such differences may be due to other causes than the change in project-selection systems, there being no way to hold all other factors constant. But it is of interest to see whether differences exist—and whether they are in accord with our hypotheses.

Finally, both hypotheses in section 6 imply that there will be an increase in firm A in the probability of commercialization (given technical completion). Why? Because firm A experienced both a closer integration between marketing and R and D and a more quantitative project-selection system. According to our hypotheses, each of these changes should increase this probability. Also, both hypotheses imply that there will be a decrease in firm B in the probability of economic success (given commercialization). Why? Because firm B experienced both a looser integration between marketing and R and D and a more quantitative project-selection system. According to our hypotheses, each of these changes should decrease this probability. Of course, these tests are not very powerful, because other factors cannot be held constant. Nonetheless, it is of interest to see whether the observed differences are in accord with these expectations.[14]

8. Empirical Results

Table 2.5 shows the effects of the reorganization of each firm on its probability of technical completion, its probability of commercialization (given technical completion), and its probability of economic success (given commercialization). The evidence is very much in accord with our hypotheses in section 6. As we would expect, the probability of commercialization (given technical completion) increased more in firm A than in firm B, and the difference is large and statistically significant.[15] With regard to the probability of technical completion and the probability of economic success (given commercialization), the differences are smaller and not statistically significant. Thus, the results suggest that a closer integration of marketing and R and D tends to increase the probability of commercialization (given technical completion). It may also have an effect on the other two probabilities, but neither the data nor our hypotheses provide any strong indication of the nature of this effect.

With regard to our second hypothesis in section 6, the evidence seems to indicate that quantitative project-selection systems tend to increase the probability of commercialization (given technical completion) and to decrease the probability of economic success (given commercialization). In accord with our hypotheses, the probability of commercialization (given technical completion) increased more in firm A than in firm C, and the probability of economic success (given commercialization) increased less in firm A than in firm C. Both of these dif-

14. When this part of the study was designed, we intended to obtain data concerning firms that experienced no change in organization or project-selection system during this period, these firms being a control group. But no suitable firms could be found, and this idea was abandoned.

15. Let Pij be the value of the probability in the ith period ($i = 1, 2$) in the jth firm ($j = a, b, c$). To test the significance of the difference between $(P_{2a} - P_{1a})$ and $(P_{2b} - P_{1b})$, we use the following statistic:

$$\frac{(P_{2a} - P_{1a}) - (P_{2b} - P_{1b})}{\sqrt{\frac{P_{1a}Q_{1a}}{N_{1a}} + \frac{P_{2a}Q_{2a}}{N_{2a}} + \frac{P_{1b}Q_{1b}}{N_{1b}} + \frac{P_{2b}Q_{2b}}{N_{2b}}}}$$

where $Q = (1 - P)$, and N is the relevant number of observations. Since the Ps—and the differences in the Ps—are statistically independent, this statistic should be distributed approximately as a unit normal variable if the null hypothesis is true.

TABLE 2.5

Probability of Technical Completion, Probability of Commercialization (Given Technical Completion), and Probability of Economic Success (Given Commercialization), before and after Reorganization, Firms A, B, and C.

PROBABILITY	BEFORE RE-ORGANIZATION (P_1)	AFTER RE-ORGANIZATION (P_2)	($P_2 - P_1$)	DIFFERENCE ASSOCIATED WITH: CLOSER INTEGRATION OF MARKETING WITH R AND D	DIFFERENCE ASSOCIATED WITH: MORE QUANTITATIVE PROJECT SELECTION SYSTEM
Economic success (given commercialization)					
A	0.55	0.75	+0.20		
B	0.39	0.57	+0.18		
C	0.30	0.75	+0.45**		
$(P_{2a} - P_{1a}) - (P_{2b} - P_{1b})$				+0.02	
$(P_{2a} - P_{1a}) - (P_{2c} - P_{1c})$					−0.25**
Commercialization (given technical completion)					
A	0.65	1.00	+0.35*		
B	0.72	0.47	−0.25		
C	0.49	0.59	+0.10		
$(P_{2a} - P_{1a}) - (P_{2b} - P_{1b})$				+0.60**	
$(P_{2a} - P_{1a}) - (P_{2c} - P_{1c})$					+0.25**
Technical completion					
A	0.61	0.65	+0.04		
B	0.68	0.54	−0.14		
C	0.52	0.63	+0.11		
$(P_{2a} - P_{1a}) - (P_{2b} - P_{1b})$				+0.018	
$(P_{2a} - P_{1a}) - (P_{2c} - P_{1c})$					−0.07

* = significant at .10 level.
** = significant at .05 level (one-tailed tests).

ferences are statistically significant. The effect of a more quantitative project-selection system on the probability of technical completion is not statistically significant. Thus, the results suggest that a more quantitative project-selection system tends to increase the probability of commercialization (given technical completion) and to reduce the probability of economic success (given commercialization), but there is no evidence that it influences the probability of technical completion.

Finally, our hypotheses imply that an increase should have occurred in firm A in the probability of commercialization (given technical completion). As shown in Table 2.5, this increase did in fact occur and is statistically significant. On the other hand, the probability of economic success (given commercialization) did not decrease in firm B, although our hypotheses imply such a decrease. However, the observed change in this probability is not statistically significant; and, as noted in the previous section, this test is not very powerful, since other variables are not held constant.

9. *Limitations and Implications*

The data presented in Table 2.5 seem to be quite consistent with our hypotheses, but since they pertain to only a few firms, and so many factors are uncontrolled, they can hardly be regarded as conclusive. Nonetheless, since this seems to be the first quantitative exploration of these issues, it may be worthwhile to point out two implications of the hypotheses in section 6, assuming that these hypotheses continue to be borne out by further study. First, the importance of integrating marketing and R and D is underlined and emphasized. The evidence presented here is quite in line with our previous results, and with Project Sappho.[16] Successful product innovation requires the coupling of marketing and technology. This coupling is difficult, particularly in a large organization where the marketing people and the R and D people tend to be separated. In cases where the marketing people are involved in R and D project selection, and where the R and D department's work is geared in considerable measure (but not entirely, of course) to marketing's perceived opportunities and needs, the productivity of R and D seems higher, and its riskiness seems lower, than in cases where there is less integration of R and D with marketing. (Of course, to the extent that more integration means a lower average return from economically successful projects, these benefits may be partially or totally offset, but there is no evidence that this is the case.)[17]

Second, formal quantitative project-selection techniques, as currently applied, seem to have at least two effects, one positive and one negative. The positive effect is that they tend to increase the probability of commercialization (given technical completion). This would be expected, since these techniques force a more careful evaluation of each project's commercial potential. The negative effect seems to be that they reduce the probability of economic success (given commercialization). To the extent that they push a firm's R and D portfolio in the direction of less ambitious projects with more modest potential payoff,[18] this too would be expected. Firms should recognize that both effects may occur, and that whether or not the positive effect outweighs the negative one is likely to differ from case to case.

16. See the references in note 12.

17. See note 1.

18. In fact, there is considerable evidence that the firms in this sample that did switch to quantitative project-selection techniques tended subsequently to reduce the proportion of their R and D portfolio aimed at relatively ambitious technical advances.

10. Underutilization of R and D Results

Finally, it is important to recognize that these probabilities of success depend heavily on how receptive other parts of the firm are to the results obtained by the R and D personnel. To get some idea of the importance of this factor, R and D executives of twenty major firms in the chemical, drug, petroleum, and electronics industries were asked to estimate what percentage of the R and D projects terminated for nontechnical reasons prior to technical completion would have been economic successes if their firm's marketing and production people had properly grasped their potential. As shown in Table 2.6, the probability of technical completion would have been about 10 percent higher, on the average, under these circumstances, according to the estimates of the eighteen firms that provided data.

Second, they were asked to estimate the percentage of technically completed projects that were not commercialized which would have been economic successes if marketing and production had properly grasped their potential. As shown in Table 2.6, the probability of commercialization (given technical completion) would have been about 26 percent higher, on the average, under these circumstances, according to the estimates. Third, they were asked to estimate the percentage of the unsuccessful commercialized projects that would have been economic successes if marketing and production had done a proper job in exploiting them. As shown in Table 2.6, the probability of economic success

TABLE **2.6**

Ratio of Value of Probability if R and D Results Were Fully Utilized to Actual Value, Eighteen Firms

	PROBABILITY OF		
RATIO	TECHNICAL COMPLETION	COMMERCIALIZATION (GIVEN TECHNICAL COMPLETION)	ECONOMIC SUCCESS (GIVEN COMMERCIALIZATION)
		(number of firms)	
1.00–1.04	12	8	5
1.05–1.09	2	2	3
1.10–1.14	3	1	0
1.15–1.19	0	0	3
1.20–1.24	0	0	0
1.25–1.29	1	3	1
1.30–1.39	0	0	1
1.40–1.49	0	1	0
1.50 and over	0	2	2
Total	*18*	*17*	*15*
Mean ratio	1.10	1.26	1.16

(given commercialization) would have been about 16 percent higher, on the average, under these circumstances, according to the estimates.

If these estimates are at all close to the mark, they indicate that the productivity of industrial R and D would be enhanced greatly by a better and fuller utilization of existing R and D findings. Under recent circumstances, about 32 percent of the R and D projects started by these eighteen firms turned out to be economic successes. According to the estimates of these executives, if marketing and production had properly utilized them, about 50 percent of the R and D projects started by these eighteen firms would have been economic successes. Thus, based on this rough measure, the success rate would have increased by over 50 percent if the results of R and D had been fully and properly utilized.

Of course, it may be objected (quite properly) that R and D executives are likely to have a biased view of the extent to which the results of R and D are underutilized. To get some idea of the probable bias, we asked a number of marketing or production executives in the same firms to answer the same questions. The results indicated that the non–R and D executives agree with the R and D executives. Indeed, their estimates of the extent to which R and D results are underutilized for these reasons tend to be somewhat higher than the estimates made by the R and D executives.

Recognizing that this problem exists, firms have tried a number of devices and strategies to promote a fuller use of R and D findings. They have tried to get marketing and production to work closely with R and D in the choice of R and D projects. They have tried to promote frequent and extensive communication between R and D and other departments. They have moved personnel back and forth across the interface between the R and D department and other departments. And they have established new-product committees and new-product departments. Unquestionably, these devices have helped a great deal. (For example, there is some evidence that, according to the estimates made by the firms in the sample, R and D results tend to be more fully utilized in firms with new-product committees.)[19] However, our results indicate that there is still a rather

19. Based on standard regression techniques, we find that the percent of a firm's technically completed, but not commercialized, R and D projects that would have been economic successes if marketing and production had properly grasped their potential (according to the R and D executives) is directly related to the percent of projects stemming from the R and D personnel and smaller if a firm has a new-products committee than otherwise. The percent of a firm's commercialized, but not economically successful, R and D projects that would have been economically successful if marketing and production had done a better job of exploiting them (according to the R and D executives) is directly related to the proportion of the firm's R and D expenditures going for "demand-pull" projects, inversely related to the firm's probability of economic success (given commercialization), and smaller if a firm has a new-products committee or a new-products department than otherwise. The percent of a firm's R and D projects that were terminated prior to technical completion for nontechnical reasons which would have been economic successes if the

serious underutilization by firms of their R and D results. Apparently, the rate of technological change could be increased significantly—without substantial increases in R and D expenditures—if firms could make fuller use of the R and D results that they are already turning out.[20]

11. Summary and Conclusions

From our findings, we may reach four general conclusions. First, there are very large differences among firms in the probability of technical completion, the probability of commercialization (given technical completion), and the probability of economic success (given commercialization) of their R and D projects. Even in the same industry, these probabilities can be much higher in one firm than in another. Among other things, these differences reflect the fact that the R and D portfolios of some firms are inherently more risky than those of others, and that some firms are more efficient than others in carrying out, managing, and utilizing R and D.

Second, based on data from sixteen major firms, there is considerable evidence that all three of these probabilities are influenced by how quickly R and D projects are evaluated from the point of view of economic potential. To the extent that firms want to increase these probabilities, they might consider an earlier investigation of the profit potential of projects. The firms with relatively high probabilities tend to be relatively quick to evaluate their economic (rather than their purely technological) potential. Also, there is considerable evidence that the probability of commercialization (given technical completion) is directly related to the degree to which R and D and marketing are integrated. It seems likely that this reflects the fact that, in some firms, the R and D personnel have worked more

---

marketing and production people had properly grasped their potential (according to the R and D executives) is directly related to the percent of projects stemming from the R and D personnel and smaller if a firm has a new-product committee or a new-product department than otherwise. All of these results are statistically significant. The sign of each of the regression coefficient seems reasonable, with one possible exception: We might have expected the proportion of expenditures going for ''demand-pull'' projects to have the opposite effect (in the second sentence in the footnote).

20. Of course, it must be recognized that R and D findings that are not put to use by a firm may be utilized by the same firm or another firm at a later date, and perhaps in a somewhat different form. It should not be assumed that these findings are lost. However, delay can be costly—both to the firm and to society.

Also, since the R and D and non–R and D executives were not asked to indicate which particular R and D findings were underutilized, it is not clear that the two groups agree on this score. All that can be determined from our results is that they seem to agree concerning the magnitude of the problem.

closely with the marketing personnel than in other firms, the result being that the R and D output has been better mated with market realities.

Third, there is evidence that the use of quantitative project-selection techniques seems to increase the probability of commercialization (given technical completion). Also, increases in the proportion of projects stemming from outside R and D seem to be associated with increases in the probability of technical completion. However, both of these factors also tend to reduce the probability of economic success (given commercialization), apparently because they push the firm's R and D portfolio in the direction of technically less ambitious projects with more modest payoffs. Whether, on balance, such changes are good or bad will vary from case to case, but it is important to recognize that such a trade off may exist. Also, there is evidence that the probabilities of technical completion and of commercialization (given technical completion) tend to be lower in firms that devote a relatively large percentage of their R and D expenditures to "technology-push" projects rather than to "demand-pull" projects. To a large extent, this may be due to the fact that such projects are inherently more risky and that other parts of the firm are less likely to accept and utilize the results of such projects.

Fourth, it is important to recognize that these probabilities depend heavily on how receptive other parts of the firm are to the firm's R and D results. According to executives (both inside and outside R and D) of eighteen major firms, the probability of technical completion would have been about 10 percent higher, on the average, if the marketing and production people had properly grasped the potential of the projects. Also, the probability of commercialization (given technical completion) would have been about 20–30 percent higher, and the probability of economic success (given commercialization) would have been about 10–20 percent higher, on the average, if marketing and production had done a proper job of exploiting them. Clearly, the productivity of industrial R and D would be greatly enhanced by a fuller utilization within the firm of its R and D results. Based on these results, it appears that the rate of technological change could be increased materially, without any substantial increase in R and D expenditures, if intrafirm barriers to the utilization of R and D results could be reduced.

In conclusion, it is very important that the limitations of this study be recognized. The data pertain to only twenty firms and, as indicated in previous sections, they are rough in many respects. Also, the hypotheses that are tested are sometimes rather crude. Clearly, the present study is only a beginning, and the results must be regarded as tentative.

3 INNOVATION, DEVELOPMENT, AND SIZE OF FIRM IN THE CHEMICAL INDUSTRY

1. Introduction

In this chapter, we analyze the relationship between innovation and size of firm in the chemical industry. A new process or product becomes an innovation when it is commercially applied or introduced for the first time. Innovation is a kep step in the process by which new technology becomes accepted. Typically, the innovator—the first firm to apply or introduce a new process or product—must bear the risks involved in trying out a new process or product. Since the innovator tends to reduce the uncertainties concerning the performance of a new technology, the innovator performs a valuable social function.

A classic issue of public policy has centered on the size of innovating firms. Schumpeter[1] argued many years ago that the very largest firms tend to carry out a disproportionately large share of the major innovations. This view is accepted by Galbraith[2] and others, who argue that modern innovations require such large outlays that only very large firms can carry them out, that R and D projects must be carried out on a large enough scale so that successes and failures can in some sense balance out, and that significant market power is required to make innovation worthwhile. Other economists have challenged this hypothesis, and have marshaled considerable evidence that sheds doubt on its validity. The purpose of this chapter is to examine the extent to which this hypothesis holds true in the chemical industry, one of the nation's most important science-based industries.

1. J. Schumpeter, *Business Cycles* (New York: McGraw-Hill, 1939).

2. J. K. Galbraith, *American Capitalism* (Boston: Houghton Mifflin, 1952).

2. Previous Results Concerning Other Industries

Some evidence concerning this issue was presented by one of the authors[3] concerning the iron and steel, bituminous coal, and petroleum industries. The results indicated that in two of the three industries, coal and petroleum, the four largest firms introduced a disproportionately large share of the innovations, while in the third, the steel industry, the largest four firms introduced a disproportionately small share of the innovations. In this context, a disproportionately large (small) share was defined to be one that was bigger (smaller) than the market share of these firms. However, when the full range of firm sizes was studied, it was found that in all three industries the biggest few firms did not do the most innovating relative to their size. Instead, the sixth largest firm seemed to do the most innovating (relative to its size) in the petroleum and coal industries, and relatively small firms seemed to do the most innovating (relative to their size) in the steel industry.

In 1971, a similar study of the ethical drug industry was presented in Mansfield, Rapoport, Schnee, Wagner, and Hamburger.[4] This study concluded that, if innovations are not weighted by importance, the four largest firms did a relatively small share of the innovating, but that if innovations are weighted by their economic importance, the share of the innovations carred out by the largest four firms was equal to their share of the market. However, when the full range of firm sizes was studied, it was found here, as in the previous three industries, that the largest few firms did not do the most innovating (relative to their size). Instead, the twelfth largest firm seemed to do the most innovating relative to its size in the pharmaceutical industry. However, in the late 1960s and early 1970s, the largest few drug firms may have increased their share of the industry's innovations. According to Grabowski and Vernon, there has been a noteworthy shift of this sort.[5]

3. E. Mansfield, *Industrial Research and Technological Innovation* (New York: Norton, for the Cowles Foundation for Research in Economics at Yale University, 1968).

4. E. Mansfield, J. Rapoport, J. Schnee, S. Wagner, and M. Hamburger, *Research and Innovation in the Modern Corporation* (New York: Norton, 1971).

5. See H. Grabowski, *Drug Regulation and Innovation* (Washington, D.C.: American Enterprise Institute, 1976). Unfortunately, one recent discussion misinterprets part of Grabowski's, Mansfield's, and Schnee's earlier results. See D. Schwartzman, *Innovation in the Pharmaceutical Industry* (Baltimore: Johns Hopkins, 1976). For a series of papers that point out some of the misunderstandings, see the comments on Schwartzman's paper by W. Comanor, H. Grabowski, and J. Schnee in J. Cooper (ed.), *Regulation, Economics, and Pharmaceutical Innovation* (Washington, D.C.: American University, 1976).

For a valuable survey which includes information concerning the sources of many chemical innovations, see American Chemical Society, *Chemistry in the Economy* (Washington, 1973).

3. Basic Data Concerning Chemical Process Innovations

To shed light on the size distribution of the innovators in the chemical industry, data are required concerning the chemical innovations that have taken place and the identity of the firms that were first to introduce each one. Let's begin by considering process innovations. (Product innovations will be considered in later sections of this chapter.) As in the previous studies, our first step was to consult various industrial and academic sources to assemble a list of important process innovations in the chemical industry. These sources included (a) chemical-engineering departments at the University of Delaware, Brooklyn Polytechnic Institute, Columbia University, and the University of Pennsylvania; (b) trade journals; (c) engineering and construction firms; and (d) chemical firms themselves. In addition to listing the processes, these sources were asked to estimate the importance of each process (based on the annual cost savings realized by replacing the old process with the new) on a scale from 1 to 10. Through trade journals and industry sources, we were able to determine which firm was the first to utilize each process commercially and approximately when this occurred.

Considerable development work frequently is required before an invention can be brought to the state of commercial applicability. (In subsequent chapters, we discuss the nature of this development work in detail.) Since the firm that is the innovator—the first to commercially introduce the innovation—may not be the developer of the innovation (for example, because the developer is a foreign firm or an engineering firm), it seemed important to look into the size of the developers, as well as the size of the innovators. Data on the firm or group that developed each innovation, as well as the innovator, were obtained from trade journals and industry sources.[6]

All of these data are presented as Table 3.1. Included is the name of each innovation, the firm that was the U.S. innovator, and the firm chiefly responsible for development of the process.[7] The data in Table 3.1 are grouped into two time periods—1930–1950 and 1951–1971. Since the chemical industry has experienced great growth and restructuring, it seems appropriate to eliminate at least a portion of these effects by considering two separate time periods. During

6. Note too that the firm or person that contributed the basic ideas or concepts underlying the innovation may be different from the developer. For example, see W. Mueller, "The Origin of the Basic Inventions Underlying du Pont's Major Product Innovations, 1920 to 1950," in National Bureau of Economic Research, *The Rate and Direction of Inventive Activity* (Princeton, Princeton University Press, 1962).

7. When more than one firm is responsible for development and/or innovation, credit is split equally among them. By a process innovation, we mean an innovation that changes the way of producing an *existing* product. Of course, product innovations often require new processes as well, but such new processes are regarded as essential parts of the product innovation and so generally *are not included in Table 3.1.* This is important to note.

TABLE 3.1

Chemical Process Innovations, 1930–1971

INNOVATION	U.S. INNOVATOR	DEVELOPER
	1930–1950	
Aniline by ammonolysis of chlorobenzene	Dow	Dow
Synthetic butanol	Union Carbide	Union Carbide
Synthetic ethanol	Union Carbide	Union Carbide
Synthetic camphor	du Pont	du Pont
Ammonia from natural gas	Shell	Uhde
Synthetic acetaldehyde	Union Carbide	Union Carbide
Synthetic urea	du Pont	du Pont
Bromine from sea water	Dow, Ethyl	Dow, Ethyl
Sodium phenolate process for hydrogen-sulfide removal	Chevron	Koppers
Type S cell for chlorine production	Hooker	Hooker
Acetic anhydride from ketene	Union Carbide	Union Carbide
Phenol from Rashig process	Durez	Rashig
Toluene recovery from petroleum	Shell	Shell
Ethylene glycol from formaldehyde	du Pont	du Pont
Butadiene from petroleum	Shell	Shell
Magnesium from sea water	Dow	Dow
Chloroform by methane chlorination	Dow	Dow
Phthalic anhydride from o-xylene	Chevron	Chevron
High-pressure nitric-acid process	du Pont	du Pont
Oxo process for alcohols	Exxon	Ruhrchemie
Acetylene from hydrocarbons	Union Carbide	BASF, Union Carbide
Butadiene from butane	Standard Oil (Cal.) Sun Oil, Phillips	Houdry, Phillips
Hydrogen cyanide by ammonia and methane	Rohm & Haas	I. G. Farben
Synthetic glycerin	Shell	Shell
Ethanol by direct hydration of ethylene	Shell	Shell
Ethyl chloride from ethane	Ethyl	Ethyl
	1951–1971	
Udex (glycol extraction) for BTX aromatics	Eastern States Petroleum	UOP, Dow
Stationary mercury cell for chlorine	Mathieson	Mathieson
Peracetic acid by acetaldehyde oxidation	Union Carbide	Union Carbide
Aniline by catalytic hydrogenation of nitrobenzene	Allied	Allied
Acetic acid by butane oxidation	Celanese	Celanese
Autoxidation route to hydrogen peroxide	du Pont	I. G. Farben
Phenol from cumene	Allied	Distillers, Hercules, Allied

TABLE **3.1** (continued)

INNOVATION	U.S. INNOVATOR	DEVELOPER
Nonelectrolytic route to carbon disulfide	Barium Reduction	Pure Oil
Gas/oil cracking to ethylene	Exxon	Exxon
Synthetic linear alcohols	Continental Oil	Ziegler, Continental Oil
Alpha olefins from cracked paraffin wax	Chevron	Chevron
Acetylene route to acrylonitrile	Monsanto	I. G. Farben
Direct air oxidation to ethylene oxide	Allied	Scientific Design
Titanium dioxide by chlorination	du Pont	du Pont
Cyclohexane-oxidation with boric acid to KA-oil	Chemstrand	Scientific Design
Stengel process for ammonium nitrate	Commercial Solvents	Commercial Solvents
Oxygen oxidation to ethylene oxide	Wyandotte	Shell
Mid-century process for aromatic acids	Amoco	Mid-Century
Sulfolane process for BTX aromatics	Shell	Shell
Propylene route to acrylonitrile	Sohio	Sohio
Acetylene by high temperature electric arc	du Pont	du Pont
Fluid-bed process for phthalic anhydride	Sherwin Williams, Reichhold, Cyanamid	Sherwin Williams, Badger, Cyanamid
Peracetic acid epoxidation of soybean oil	Union Carbide	Union Carbide
Fractionation of ethylbenzene from petroleum	Cosden	Cosden, Badger
Dealkylation of toluene to benzene	Signal Oil	Ashland, UOP
Hydrodealkylation route to napthalene	Signal Oil	Ashland, UOP
Acetaldehyde by oxidation of ethylene	Celanese	Wacker
Trichloroethylene from ethylene	Dow, PPG	Dow, PPG
Vinyl acetate by vapor phase oxidation of ethylene	National Distillers	Bayer, Hoechst, National Distillers
Vinyl chloride by oxychlorination process	Goodrich	Goodrich
Large-scale ammonia plants	Monsanto	Kellogg
Isoprene from propylene	Goodyear	Scientific Design, Goodyear
Methanol by low-pressure process	Monsanto	ICI
Synthesis gas by partial oxidation of hydrocarbons	Spencer	Texaco, Hydrocarbon Research
Glycerine by acrolein route	Shell	Shell
Thermal dealkylation of toluene to benzene	Crown Central Petroleum	Houdry
Acetic acid by methanol carbonylation	Monsanto	Monsanto
TPA-ethylene glycol process for polyester	Chemstrand	Calico Printers, du Pont
Propylene oxide by direct oxidation	Oxirane	Halcon
Urea process for melamine	Allied	Allied, Cyanamid

TABLE **3.1** (continued)

INNOVATION	U.S. INNOVATOR	DEVELOPER
Ethylene polymerization to alphaolefins	Gulf	Gulf
Chloroprene by butadiene oxidation	Petro-Tex	Distillers
Acrylonitrile hydrodimerization to adiponitrile	Monsanto	Monsanto
BASF process for phthalic anhydride	BASF, Monsanto, Koppers	BASF
Acrylates by vapor-phase oxidation of propylene	Union Carbide	Union Carbide, BP
One-step process to ethyleneimine	Dow	Dow

1930–1950, the chemical industry grew enormously. Many of the latter-day giants of the industry were quite small (as measured by assets or sales) in 1930,[8] but, due in part to rapid technical change, they grew phenomenally during the subsequent twenty years. The period 1951–1971 was one that brought two important structural changes to the industry. First, the development of petrochemicals and efficient methods for their production brought greatly expanded markets and profit possibilities to the industry. It also brought the petroleum firms into the chemical business on a large scale. The advent of the oil companies led to the second major structural shift—the development of large-scale production techniques. These two factors—development of petrochemicals and use of new large-scale plant techniques—revolutionized the chemical business.

Having obtained the data in Table 3.1, we next obtained measures of firm size for the two periods. The measures used were (a) asset size (in tens of millions of dollars) in 1930 for period 1 (1930–1950), and (b) asset size (in tens of millions of dollars) in 1950[9] for period 2 (1951–1971). These data were obtained from *Moody's Industrials Manual* for the appropriate years. The chemical industry, for purposes of the analysis of this section, was defined as all those domestic firms which were listed under "chemicals" in the industry-report section of *Moody's Industrials Manual,* except for those whose business was not primarily chemical, e.g., petroleum and drug firms. The impact of the latter firms will be considered later in this chapter.

8. For example, in 1930, the assets of Dow Chemical were $21.5 million, and Monsanto $23.9 million. By 1950, they had grown to $327.2 million and $249.8 million respectively.

9. Originally, gross sales was selected as the size measure. However, in 1930 a very common practice of chemical firms was not to report sales figures. Consequently, asset figures had to be used. To maintain comparability, it was necessary to use asset figures in period 2 as well.

4. The Largest Four Firms' Share of the Process Innovations

To begin with, we follow the procedure of earlier studies and compare the share of innovations contributed by the largest four firms with their share of the industry (which is defined as the percentage of total assets of the industry accounted for by the top four firms). One would expect the industry share of the largest firms to equal their share of the innovations if the largest firms allocate the same proportion of their resources to inventive activity as do other firms, if they can obtain applicable results as easily, and if they can apply results as quickly.

In period 1, the proportion of assets accounted for by the top four firms was 67 percent; and in period 2, it was 57 percent. These figures are presented in Table 3.2 along with the corresponding figures for (a) the share of the innovations of each period accounted for by the four largest firms and (b) the share of the developments of each period accounted for by the four largest firms. The percent of developments and of innovations is presented in two forms. The unweighted column counts all innovations equally and thus does not take into account the relative importance of different innovations.[10] The weighted measure, on the other hand, tries to discriminate among innovations on the basis of importance. The weights used in developing Table 3.2 were the average index of importance (on a scale of 1 to 10) obtained from the interviews discussed above.[11]

Table 3.2 suggests several tentative conclusions, admittedly crude, about the role of the largest firms in the chemical industry. First, on either a weighted or an

TABLE 3.2

Percentage of Chemical Process Developments, Innovations, and Assets Accounted for by the Four Largest Chemical Firms, 1930–1950 and 1951–1971

	UNWEIGHTED	WEIGHTED
	(percent)	
1930–1950		
Process developments	65	63
Process innovations	58	56
Assets	67	67
1951–1971		
Process developments	59	58
Process innovations	43	41
Assets	57	57

10. Unweighted data have obvious deficiencies. For example, it may be that some technical changes in Table 3.1 could be regarded as sets of innovations. If so, the unweighted results would, of course, change.

11. Although the weights may eliminate some of the problems of an unweighted analysis, many problems still remain. See Mansfield, *Industrial Research, op. cit.*

unweighted basis, the Schumpeterean argument does not seem valid with respect to process innovations in the chemical industry. In both periods, the asset share of the largest four firms exceeded their share of innovations. Second, on either a weighted or an unweighted basis, the largest firms seemed to account for a larger share of process developments than of process innovations. In both time periods, the largest firms accounted for about the same share of the process developments as they did of the industry's assets. However, as in other such comparisons, it is important to avoid the appearance of spurious accuracy: clearly, these figures are only very crude indicators.

Fourth, when one compares these results with those of earlier studies, it appears that, relative to their industry shares, the four largest petroleum and coal firms have done more innovating with respect to processes than have the four largest chemical firms. During 1939–1958, the four largest petroleum firms, with about 39 percent of the industry's capacity, accounted for about 58 percent of the process innovations; and the four largest coal firms, with about 13 percent of the industry's output, accounted for about 30 percent of the process innovations. On the other hand, relative to their share of the industry, the largest four steel firms have done less innovating than the largest four chemical firms. Finally, it is important to recall from note 7 that only new processes for producing *existing* products are included here.

5. A More Complete Analysis

The previous section provided a crude estimate of the extent to which the largest four firms in the chemical industry carried out process developments and innovations. Our analysis now shifts to include the full range of firm sizes in order to estimate which firms have done the most innovating relative to their size. Specifically, we assume that:

$$N_j = a_0 + a_1 S_j + a_2 S_j^2 + a_3 S_j^3 + z_j, \tag{3.1}$$

where N_j is the number of developments or innovations (weighted or unweighted) carried out by the jth firm, S_j is the asset size (in tens of millions of dollars) of the jth firm, and z_j is a random error term. The relation specified in equation (3.1) allows us to estimate which firms have done the most innovating relative to their size, since once the relation is estimated, one can solve for N_j/S_j, and find its maximum value.[12]

12. In estimating the model, terms that were not significant at the 5 percent level were dropped and the model was re-estimated. A few of the firms in Table 3.1 could not be included because their size could not be determined, because they were not in existence in 1930 or 1950, or because they were so closely associated with a foreign firm that their domestic size is misleading.

Table 3.3 presents least-squares estimates of the as for both developments and innovations, weighted and unweighted, for both time periods studied. The numbers in parentheses are t values, and the coefficient of determination is adjusted for degrees of freedom. In period 1, for both weighted and unweighted data, higher-order terms did not provide a statistically superior relationship, so results for the linear model are reported. Since the estimated value of a_0 in each case was not significantly different from zero, it was omitted. Given the form of the relationship, it is clear that, on the average, N_j/S_j was the same for all relevant sizes of firm. Thus, in period 1, there seemed to be no tendency for large firms to carry out any more process developments or innovations (relative to their size) than small firms.

Let's turn now to period 2. The quadratic form of equation (3.1) is statistically superior to the other forms in period 2. Regardless of whether one uses unweighted or weighted data, the estimate of a_0 is not statistically significant for

TABLE **3.3**

Least-Squares Estimates of a_0, a_1, a_2, and a_3, Chemical Process Developments and Innovations, 1930–1950 and 1951–1971

	a_0	a_1	a_2	a_3	$\overline{R}^2$	N
		1930—1950				
Number of developments						
Unweighted	—[a]	0.059 (6.044)	—[a]	—[a]	.45	38
Weighted	—[a]	0.357 (5.206)	—[a]	—[a]	.37	38
Number of innovations						
Unweighted	—[a]	0.062 (6.044)	—[a]	—[a]	.45	38
Weighted	—[a]	0.372 (5.175)	—[a]	—[a]	.36	38
		1951–1971				
Number of developments						
Unweighted	−0.090 (1.978)	0.0527 (10.89)	−0.00024 (6.477)	—[a]	.81	56
Weighted	−0.563 (1.925)	0.327 (10.574)	−0.00168 (7.105)	—[a]	.77	56
Number of innovations						
Unweighted	—[a]	0.0822 (6.544)	−0.000430 (4.224)	—[a]	.48	56
Weighted	—[a]	0.533 (6.260)	−0.00309 (4.472)	—[a]	.42	56

[a] Omitted because not statistically significant.

innovations, but significant for developments. Given the form of the resulting relationships, it is clear that the maximum value of N_j/S_j occurs among very small firms for innovations (regardless of whether weighted or unweighted data are used). For developments, the maximum value of N_j/S_j occurs at a 1950 asset level of about $200 million, which was about the size of the seventh or eighth largest firm. This was true for both the weighted and unweighted data. Thus, there seemed to be no tendency for the very largest firms to carry out a disproportionately large share of the process innovations or developments in the later period either.[13]

6. *Sources of Chemical Process Developments and Innovations*

Up to this point, we have considered the behavior of only those firms whose primary business was chemical. A further question of interest is the extent to which firms other than traditional chemical firms have accounted for process developments and/or innovations. It is well known that many petroleum firms, rubber firms, paint firms, and other firms have very substantial chemical operations. To what extent have these firms been responsible for major developments and innovations? Also, many new processes may have been developed by foreign firms or by engineering firms. To what extent have these firms been important sources of significant developments, and how has their importance varied over time?

In this section, we look at the sources of process development and innovation.[14] These sources are classified into five separate groups: (1) the group of firms whose business is primarily chemical (i.e., the firms we considered in the prior sections), (2) petroleum firms, (3) foreign firms, (4) engineering and construction firms, and (5) other firms. The last category includes firms that are primarily engaged in making rubber, paint, etc.—companies that do not fit into the previous categories. Then we computed for each time period, for developments and innovations (both weighted and unweighted), the proportion of devel-

13. In our previous studies, we have often looked at the ratio of the estimated value of N_j in the later period to that in the earlier period, holding constant S_j. However, there is such a considerable range of the size distribution in the later period that exceeds the size of the biggest firm in the earlier period that we decided not to make such calculations in this case.

14. Of course, the establishment of a single source for a chemical process development or innovation is often hazardous. Often an innovation depends on a set of separate developments. For example, a new process may only become feasible when a better catalyst is developed. Moreover, the innovation often is the result of a growing amount of knowledge with each part essential to the successful application of the process. The same is true of new products. Also one publication or expert sometimes gives much more credit to a given source than another publication or expert. Thus, the data in Table 3.1 (and Table 3.5) should be viewed with caution.

opments or innovations in Table 3.1 accounted for by each category of firm. The results of this analysis are shown in Table 3.4.

These results seem to indicate several things. First, during 1951–1971, about three fifths of the process innovations were carried out by chemical firms, about one quarter were carried out by petroleum firms, and about one tenth were carried out by other domestic nonengineering firms. Thus, firms whose primary

TABLE **3.4**

Percentage of Chemical Process Developments and Innovations Arising from Various Sources, 1930–1950 and 1951–1971[a]

SOURCE	UNWEIGHTED	WEIGHTED
1930–1950		
	(*percentage*)	
Developments		
Chemical firms	50	49
Petroleum firms	21	23
Foreign firms	17	18
Engineering firms	2	2
Other firms	10	8
Total	*100*	*100*
Innovations		
Chemical firms	60	58
Petroleum firms	35	36
Foreign firms	0	0
Engineering firms	0	0
Other firms	6	6
Total	*100*	*100*
1951–1971		
Developments		
Chemical firms	33	31
Petroleum firms	25	24
Foreign firms	18	20
Engineering firms	18	18
Other firms	6	7
Total	*100*	*100*
Innovations		
Chemical firms	58	58
Petroleum firms	30	27
Foreign firms	1	1
Engineering firms	1	2
Other firms	10	11
Total	*100*	*100*

[a] Because of rounding errors, the individual figures may not always sum to the totals.

business is nonchemical have accounted for a substantial share of the chemical process innovations. Moreover, their share of process developments was even greater than their share of process innovations. In 1951–1971, the chemical firms accounted for only about one third of the process developments, while petroleum firms accounted for about one quarter, engineering firms accounted for about 18 percent, and other domestic nonengineering firms accounted for about 7 percent.

Second, while the traditional chemical firms have maintained their share of process innovations, their share of process developments was substantially lower in 1951–1971 than in 1930–1950. Specifically, they accounted for about one half of the process developments in 1930–1950, as contrasted with about one third in 1951–1971. The engineering and construction firms have increased their share of process developments spectacularly in this period. Specifically, they accounted for about 18 percent of the process developments in 1951–1971, as contrasted with about 2 percent in 1930–1950. The increase in the engineering and construction firms' share seems to have been at the expense of the chemical firms' share.

Third, the share of chemical process innovations carried out by "other firms" seems to have increased considerably. In 1951–1971, such firms accounted for about 11 percent of these innovations, whereas in 1930–1950 they accounted for about 6 percent. This, of course, reflects the fact that so many nonchemical firms entered the chemical business in recent years. The increase in their share of process innovations seems to have been at the expense of the petroleum firms' share. In 1930–1950, the petroleum firms accounted for about 36 percent of chemical process innovations, as contrasted with about 27 percent in 1951–1971.

Finally, these very crude figures suggest that foreign firms have accounted for about one-fifth of the developments underlying U.S. chemical innovations. (For some warnings concerning this result, see note 23.) Naturally, the foreign firms' share of innovations has been very small throughout the relevant period.

7. *Basic Data Concerning Chemical Product Innovations*

Let's turn now to product innovations in the chemical industry. As in the case of process innovations, various industry sources were asked to list the most important product innovations during the period 1930–1966.[15] As before, these

15. The later period for product innovations ends with 1966 rather than 1971, because it was felt that it is particularly difficult to estimate the importance of a relatively new product innovation. Note too that in some cases these product innovations may represent the first large-scale production of these products, there having been some small-scale production before.

sources included chemical-engineering departments at universities, the staffs of various trade journals, the staffs of chemical consulting firms, and personnel in chemical firms themselves. From trade journals and further discussions with industry sources, the following information was gathered for each new product: (1) the first U.S. firm to commercially produce the new product, (2) the firm or group most responsible for developing the product in order to bring it to a state of commercial feasibility, and (3) an estimate of the importance of the innovation (as measured by its sales volume) on a scale from 1 to 10. This latter figure will be used in subsequent analyses as a measure of the economic importance of individual products, and so corresponds directly to the weights used for process innovations.

The data are presented in Table 3.5. It is important to recognize that sales volume is used as an index of importance. To illustrate some of the problems in this index, note that some of the new products in Table 3.5 have been banned or restricted in recent years. Since the number of such cases is relatively small, they may not have much quantitative impact on our findings. But this does not alter the fact that sales volume can be a very crude measure of importance. Unfortunately, it is difficult to find better alternative measures that are feasible in work of this kind.

8. The Largest Four Firms' Share of the Product Innovations

Using the data in Table 3.5, we analyze the extent to which the four largest chemical firms developed and innovated new products in the chemical industry. The four largest chemical firms accounted for 67 percent of the industry's assets in period 1 and 57 percent in period 2.[16] Under the assumption that the biggest firms allocate the same proportion of their resources to developmental and innovative activity as do other firms, that they can obtain commercially viable results as easily, and that they can apply these results as easily, one would expect their share of developments and innovations to be approximately equal to their share of assets. Table 3.6 shows their share of developments, innovations, and assets both for period 1 and period 2. The shares of new-product developments and new-product innovations are presented in two forms. The first is an unweighted form that considers all developments and innovations as equally important. The second, or weighted, set of data weights each development (or innova-

16. Recall that the industry, as defined here, is composed of all those firms listed as chemical firms in *Moody's Industrials Manual* in 1930 for period 1, and 1950 for period 2, except that firms engaged primarily in some other business and foreign firms are omitted. Also, note that, in some cases, the innovator here is different from that in our study of the petroleum industry cited in note 3. This is because the first *oil* company to produce the new product was considered the innovator in the latter study.

TABLE 3.5

Chemical Product Innovations, 1930–1966

INNOVATION	U.S. INNOVATOR	DEVELOPER
	1930–1950	
Maleic acid	Allied	Allied
Sodium chlorite in textone process	Mathieson	Mathieson
Polysulfide rubber	Thiokol	Thiokol
Maleic anhydride	Allied	Allied
Methyl and ethyl acrylates	Rohm & Haas	Rohm & Haas
Cellulose acetate butyrate	Kodak	Kodak
Neoprene	du Pont	du Pont
Freon	du Pont	General Motors
Polyvinyl chloride	Union Carbide	I. G. Farben
Cordura high-tenacity rayon	du Pont	du Pont
Ethyl cellulose	Hercules	Lilienfeld, Leuchs, Dreyfus
Perchloroethylene	du Pont	du Pont
Urea-formaldehyde resins	Cyanamid	British Cyanide, Plaskon
Propylene glycol	Union Carbide	Union Carbide
N-butanol and butyraldehyde	Union Carbide	Union Carbide
Glass fibers	Corning, Owens Illinois	Corning, Owens Illinois
Sorbitol and mannitol	Atlas	Atlas
Polyvinyl butyral	Shawinigan and Monsanto	Shawinigan
Nitroparaffins	Commercial Solvents	Commercial Solvents, Purdue
Polyvinyl alcohol	du Pont	Consortium für Elektro-Chemische Industrien
Methyl methacrylates	Rohm & Haas, du Pont	ICI, Rohm & Haas
Polystyrene	Dow, U.S. Rubber	I. G. Farben, Dow, U.S. Rubber
Methyl cellulose	Dow	Dow
Nylon 6/6	du Pont	du Pont
Melamine-formaldehyde resins	Cyanamid	Henkel, CIBA, Cyanamid
Acrylonitrile	Cyanamid	Mouseu
Vinylidene chloride homo- and co-polymers	Dow	Dow
High-pressure polyethylene	du Pont, Union Carbide	ICI
Butyl rubber	Exxon	Exxon
Cyclohexane	du Pont	Baeyer, Sabatier
Polyester resins	PPG	PPG, Cyanamid, Plaskon
Teflon	du Pont	du Pont
Silicon polymers	Corning Glass, Dow	Corning Glass, GE
DDT	Cincinnati Chemical Works	Geigy
Polyvinyl acetate	Union Carbide	Consortium für Elektro-Chemische Industrien

TABLE **3.5** (continued)

INNOVATION	U.S. INNOVATOR	DEVELOPER
Alkylbenzene sulfonates	Allied	I. G. Farben, Allied
Kel-F	Kellogg	Kellogg
2, 4, 5–T	Amchem	Amchem, N.Y. State Agricultural Experiment Station
A-B-S resins	U.S. Rubber	U.S. Rubber
Orthoxylene	Standard Oil (Cal.)	Standard Oil (Cal.)
Epoxy resins	Devoe & Reynolds	CIBA, Shell, Devoe & Reynolds
Benzene hexachloride	Hooker	ICI
2, 4-D	Amchem	N.Y. State Agricultural Experiment Station, Bryce Thompson, Amchem
P-S-B resins	Dow	Dow
Parathion	Cyanamid	Bayer
PSA (SAN) resins	Rohm & Haas	Dow
Modacrylics	Union Carbide	Union Carbide
Isoprene	Newport Industries	Newport Industries
Allyl chloride	Shell	Shell
Titanium	du Pont, National Lead	Kroll, Bureau of Mines
Acrylic fiber	du Pont	du Pont
Cyclamates	Abbott	du Pont
	1951–1966	
Chlorosulfonated polyethylene	du Pont	du Pont
Para-xylene	Standard Oil (Cal.)	Standard Oil (Cal.)
Triazine herbicides	Geigy	Geigy
Polyester fiber	du Pont	du Pont, ICI, Calico Printers
Captan	Standard Oil (Cal.)	Standard Oil
Ethylene/propylene rubber	Exxon	du Pont, Natta, Exxon Hercules, Ziegler, U.S. Rubber, Avisun
Acetal resins	du Pont	du Pont
Isocyanates	Bayer, Monsanto, du Pont	Bayer, du Pont
Low-pressure polyethylene	Phillips, Grace, Celanese, Union Carbide, Hercules	Ziegler, Phillips, Standard Oil (Ind.)
Trimellitic anhydride	Standard Oil (Ind.)	Standard Oil (Ind.)
Polycarbonate resins	Monsanto, Bayer	GE, Bayer
Polypropylene	Hercules	Montecatini, Ziegler, Natta
Polyisoprene rubber	Shell	Shell
Carbamates (Sevin)	Union Carbide	Union Carbide
Thiocarbamate-based herbicides	Stauffer	Stauffer

TABLE 3.5 (continued)

INNOVATION	U.S. INNOVATOR	DEVELOPER
Nylon 6	Allied, American Enka	I. G. Farben
Tetramethyl lead	Ethyl	Ethyl
Treflan	Eli Lilly	Eli Lilly
Isopthalic acid	Standard Oil (Cal.)	Standard Oil (Cal.)
Polybutadiene rubber	Phillips	Phillips
Malathion	Cyanamid	Cynamid
Polyvinylidene fluoride	Penn Salt	Penn Salt
Phenoxy resins	Union Carbide	Union Carbide
Polyallomers	Kodak	Kodak
Polyimides	du Pont	du Pont
Polyphenylene oxide	GE	GE, AKU
Ionomers	du Pont	du Pont
Polysulfones	Union Carbide	Union Carbide
Parylene	Union Carbide	Union Carbide
Chloroacetamide pre-emergence pesticides	Monsanto	Monsanto

tion) by the average index of economic importance (described above) obtained for it in the interviews.

Though a much simplified and crude measure of the role of the largest four firms, Table 3.6 points to some interesting conclusions. First, the results for 1930–1950 indicate that the largest four firms carried out a smaller share of the product innovations and developments than their share of the industry's assets. Second, turning from 1930–1950 to 1951–1966, Table 3.6 indicates that the largest four firms carried out a larger share of the product innovations and devel-

TABLE 3.6

Percentage of Chemical Product Developments, Innovations, and Assets Accounted for by the Four Largest Chemical Firms, 1930–1950 and 1951–1966

	UNWEIGHTED	WEIGHTED
	(*percent*)	
1930–1950		
Product developments	53	55
Product innovations	61	63
Assets	67	67
1951–1966		
Product developments	69	70
Product innovations	61	60
Assets	57	57

opments than their share of the industry's assets. Third, comparing these results with earlier results for other industries, it appears that the largest four firms in the chemical industry have done much more product innovating (relative to their size) than the largest four firms in the steel industry, about the same amount (relative to their size) as the largest four firms in the petroleum industry, and somewhat more (relative to their size) than the four largest ethical drug firms (at least up to the early 1960s).[17]

9. *A More Complete Analysis*

Let's look now at the full range of firm size, not just the four largest firms, to assess the importance of each firm size relative to all other sizes. The model assumed for this analysis corresponds directly to that employed in section 5. Specifically, we assume that:

$$N_j = a_0 + a_1 S_j + a_2 S_j^2 + a_3 S_j^3 + Z_j, \qquad (3.2)$$

where, as before, N_j is the number of developments or innovations (weighted or unweighted) carried out by the *j*th firm, S_j is the asset size of the *j*th firm (in tens of millions of dollars), and Z_j is a random error term. Once this relation is estimated, it can, of course, be manipulated to find the value of S_j where the number of developments or innovations is biggest, relative to firm size—i.e., where N_j/S_j is a maximum.[18]

Table 3.7 presents the least-squares estimates of the *a*s based on the data in Table 3.5. Results are given for developments and innovations, unweighted and weighted, for period 1 and period 2. The numbers in parentheses are *t* statistics, the coefficients of determination are adjusted for degrees of freedom, and *N* is the number of firms in the industry. (For the definition of the industry, see section 3 above.) For innovations, according to the results, the maximum value of N_j/S_j occurred at the size of the biggest firm, du Pont. This was true for both the

17. See the studies cited in notes 3 and 4. As a summary figure for chemicals, we use here the average of the figure for developments and the figure for innovations. The results indicate that the average percentage for the four largest chemical firms was somewhat higher than their share of assets. The figure for steel is lower, whereas those for petroleum and drugs are about the same, as their share of sales. Note, however, that the results here are based on assets but that the results in the other industries are based on sales. But since the largest firms' share of assets probably exceeds their share of sales, the results would probably be the same if sales were used in all industries. Also, note that the results obtained by Grabowski and Vernon suggest that the largest four drug firms did much more innovating (relative to their size) after the mid-1960s than before.

18. As before, whenever terms are not significant at the 5 percent level, they are excluded. Also, some firms had to be omitted for the reasons given in note 12.

TABLE 3.7

Least-Squares Estimates of a_0, a_1, a_2, and a_3, Chemical Product Developments and Innovations, 1930–1950 and 1951–1966

	a_0	a_1	a_2	a_3	R^2	N
			1930–1950			
Number of developments						
Unweighted	—[a]	—[a]	0.0018 (10.19)	—[a]	0.70	38
Weighted	—[a]	—[a]	0.0122 (10.13)	—[a]	0.70	38
Number of innovations						
Unweighted	—[a]	0.2193 (3.181)	−0.00603 (1.81)	0.0000842 (2.166)	0.78	38
Weighted	—[a]	0.607 (3.164)	—[a]	0.000149 (2.306)	0.78	38
			1951–1966			
Number of developments						
Unweighted	—[a]	−0.0131 (2.108)	0.00110 (6.776)	−0.00000525 (6.037)	0.92	56
Weighted	—[a]	—[a]	0.00299 (9.601)	−0.0000106 (4.799)	0.95	56
Number of innovations						
Unweighted	—[a]	—[a]	0.000909 (10.155)	−0.00000444 (7.002)	0.89	56
Weighted	—[a]	0.113 (4.310)	0.000875 (4.110)	—[a]	0.89	56

[a] Omitted because not statistically significant.

weighted and unweighted data and for both 1930–1950 and 1951–1966.[19] Note that this does not contradict our finding in the previous section that the largest four chemical firms accounted for a smaller share of the innovations in 1930–1950 than they did of the industry's assets. Whereas this was true of the largest four firms taken as a whole, it was by no means true of du Pont. In other words, du Pont did more than its share of the innovating, but some of the other members of the "top four" did not.

The results for product developments are essentially the same as for product

19. In the case of unweighted innovations in 1930–1950, there is a double maximum, one occurring among the smallest firms and one at the largest firm; but when the innovations are weighted by importance, there is but a single maximum—at the size of the largest firm. Also, in the case of unweighted innovations in 1951–1966, the maximum occurs between the size of the first and second largest firms. However, when the data are weighted, it occurs at the size of the largest firm.

innovations. In 1930–1950, the biggest firm, du Pont, seemed to carry out the most product developments relative to its size. This is true for the weighted and unweighted data. Again in 1951–1966, the maximum value of N_j/S_j is reached at the size of the largest firm, du Pont. And this is true for both the weighted and the unweighted data.[20]

10. Sources of New Product Developments and Innovations

In section 6, we saw that many sources other than the traditional domestic chemical firms—that is, firms whose business is primarily the manufacture of chemicals—have been responsible for process innovations. In this section, we look at the sources of important product innovations, these sources again being classified as (1) chemical firms—i.e., the firms used in the analysis of the prior sections; (2) petroleum firms; (3) foreign firms; (4) engineering and construction firms; and (5) other firms.[21]

Table 3.8 shows, for each time period, and for weighted and unweighted data, the proportion of developments and innovations accounted for by each group. Several noteworthy findings arise from this table. First, although the chemical firms introduced about three fifths of the product innovations in period 2 and about three fourths in period 1, a substantial proportion of these innovations were developed by others, particularly foreign firms. During 1951–1966, only about two fifths of the weighted developments were carried out by domestic chemical firms.

Second, with the full-scale entry of petroleum firms into the chemical business in the 1950s and 1960s, the petroleum firms were responsible for a much larger share of the innovations and developments in period 2 than in period 1. In period 1, they accounted for a mere 6 percent of innovations and 5 percent of developments on a weighted basis. However, in period 2, their share of both developments and innovations rose to 24 percent on a weighted basis.

Third, in contrast to the results for process innovations, the engineering and construction firms are quite unimportant sources of new product developments and innovations. This seems reasonable since the business of the engineering and construction firms is to sell plant and equipment which embodies new and better process technology. However, this conclusion must be viewed with some caution. Quite often, the commercial production of a new chemical product requires that one develop an entirely new commercial process for making the product.

20. More precisely, the maximum occurs between the size of the first and second largest firms in the case of the unweighted data in 1951–1966.

21. As before, this "other" group contains primarily such firms as rubber, paint, and glass companies.

TABLE **3.8**

Percentage of Chemical Product Developments and Innovations Arising from Various Sources, 1930–1950 and 1951–1966[a]

SOURCE	UNWEIGHTED	WEIGHTED
	(percentage)	
1930–1950		
Developments		
Chemical firms	48	45
Petroleum firms	6	6
Foreign firms	27	29
Engineering firms	2	1
Other firms	12	14
Universities, research institutes, and government agencies	5	5
Total	*100*	*100*
Innovations		
Chemical firms	73	73
Petroleum firms	6	5
Foreign firms	1	1
Engineering firms	2	1
Other firms	18	20
Total	*100*	*100*
1951–1966		
Developments		
Chemical firms	44	42
Petroleum firms	23	24
Foreign firms	19	23
Engineering firms	0	0
Other firms	14	10
Total	*100*	*100*
Innovations		
Chemical firms	56	58
Petroleum firms	24	24
Foreign firms	6	7
Engineering firms	0	0
Other firms	14	10
Total	*100*	*100*

[a] Because of rounding errors, individual figures may not sum to totals.

Any such new processes are not included in our study as separate innovations but are included as part of the product innovation. Thus, to the extent that engineering and construction firms aided in the development of such new processes, their importance in the development of new products may be understated.

Fourth, foreign firms have been a very important source of new product developments. In both periods, they accounted for 20–30 percent of weighted developments, and (for reasons discussed in note 23) this may be an understatement. Foreign firms have been a much less significant source of innovations. Apparently, foreign firms that developed a new product have generally licensed a domestic firm to produce it in the United States. An excellent example is low-density polyethylene, which ICI licensed du Pont to produce in the United States. However, as shown in Table 3.8, the foreign firms' share of innovations in the United States seems to be increasing.

11. Comparison of Results for Process Innovations and Product Innovations

Before concluding this chapter, it is worthwhile to compare the results for process innovations with those for product innovations. Clearly, there are some important differences between these results. First, the role of the biggest firms is much greater for product innovations than for process innovations. In the case of processes, there was no evidence in either time period that the largest firms carried out more innovations or developments than would have been expected on the basis of their share of the industry's assets. But in the case of products, the greatest number of innovations and developments (relative to firm size) was reached at the size of the largest firm, du Pont. To a considerable extent, this difference undoubtedly reflects a conscious policy decision at du Pont concerning the areas and types of work that it regards as most profitable. New products like nylon or orlon can be enormously profitable if a firm is able to overcome the many technical and commercial hurdles that lie along the road to success.

Second, the sources of process innovations and developments have been somewhat different than those of product innovations and developments. In 1930–1950, petroleum firms played a much more important role in developing and innovating chemical processes than in developing and innovating chemical products. However, during the 1950s and 1960s, the petroleum firms' role with regard to product development and innovation increased to the point where it equaled their role with regard to processes. Another important difference is that, during the 1950s and 1960s, engineering and construction firms have been a much more important source of process developments than of product developments. The reasons for this were noted in a previous section.

While stressing the importance of these differences between processes and products, it is important at the same time to recognize some major similarities between them. First, both for processes and products, the relationship between a firm's innovative or developmental performance and its size is by no means so close or dependable that one can predict a firm's performance very well from its size alone. On the contrary, there are some small firms—like Dow in the early period—that seem to have done as much innovating and developing (relative to their size) as any firm, both with regard to processes and products. Moreover, there are some very large firms that seem to have done very little innovating and developing, both with regard to processes and products. Although the average relationships between firm size and innovative or developmental performance are of interest, they should not be regarded as showing anything more than relatively crude averages about which there is considerable dispersion.

Second, both for processes and products, innovations were not developed by the innovator in a substantial proportion of cases. Moreover, both for processes and products, the probability that the U.S. innovator is also the developer tends to increase with the size of the innovator. In other words, large firms tend to be more likely to have developed the innovations that they introduce. There are several reasons for this. Smaller firms may be more inclined to license a new process or product from another firm, since they have less resources to devote to R and D. Also, the licensor may be more inclined to deal with the smaller firm because it may be able to extract a more favorable licensing and royalty arrangement than it would from a very big firm. Furthermore, when one firm announces a development, other firms—and particularly the largest ones—are likely to have been working along similar lines, and they may prefer to continue to develop their own version of the innovation, which they hope will have somewhat better properties or lower costs. Thus, the most ready market for the developer, if he chooses to license rather than to introduce the innovation himself, may be a smaller firm.

Incidentally, these results tend to put Mueller's findings regarding du Pont's innovations in a somewhat different perspective.[22] Mueller showed that a very large share of du Pont's most important product and process innovations were based on inventions made outside of du Pont. Based on the present study, it is clear that du Pont is not the only chemical firm of which this is true. On the contrary, when compared with other American chemical firms, du Pont seems to have developed a larger share of its own innovations than its smaller competitors.

Finally, another similarity between our results concerning process and product innovations is that, in both cases about 20 to 30 percent of American innovations were based on developments by foreign firms (or other kinds of foreign organizations). In many of the cases where the American innovator did not develop the innovation, the development came from abroad. Thus, both for products and processes, the United States makes considerable use of foreign technology, particularly that originating in the United Kingdom and West Germany. However, the specific percentages given in this regard should be viewed with particular caution.[23]

22. See Mueller, *op. cit.*

23. It is interesting to compare this figure with the National Science Foundation's finding that about 65 percent of the major technological innovations in 1953–1973 were

12. Summary and Conclusions

For many years, economists have been interested in the role of the biggest firms in the innovative process and in the sources of major innovations. In this study, we have looked in some detail at the innovative performance of firms of various sizes in the chemical industry, and at the sources of chemical innovations. Our results can be summarized as follows: First, in the case of processes, there is no evidence in either 1930–1950 or 1951–1971 that large chemical firms accounted for more developments and innovations (relative to their size) than smaller ones. However, in the case of products, the largest firm, du Pont, seemed to carry out the most developments and innovations (relative to its size), both in 1930–1950 and 1950–1966.

Second, comparing these results with those obtained before for the steel, petroleum, coal, and pharmaceutical industries, we find that this is the first case encountered to date where the biggest firm in an industry has done the most innovating (relative to its size). This result is in close accord with previous findings, which indicate that the chemical industry is the one major industry where the biggest firm spends more (relative to its size) on R and D than somewhat smaller firms. Also, du Pont seems to be regarded by other firms in the industry, and by knowledgeable observers outside the industry, as having the industry's most productive R and D establishment.[24]

Third, whether du Pont has been so technologically productive partly because of its size or whether its size has been a handicap that it has overcome is a question that cannot be answered on the basis of our findings. But our results indicate clearly that large size does not insure technological progressiveness in the chemical industry. Some large chemical firms produced relatively few innovations, whereas, in the past at least, some small firms carried out a great many. In partic-

---

produced by the United States, while 35 percent came from abroad. See National Science Board, *Science Indicators, 1974,* (Washington, D.C.: Government Printing Office, 1975). The latter percentage, which pertains to all industries, not just chemicals, is not too different from the 20 to 30 percent cited in the text.

At the same time, it is important to stress the crudeness of our figures. For one thing, some of the data may have been biased in favor of domestic sources. (Thus, some authorities credit the development of 2,4-D to ICI, not the sources given in Table 3.5.) The OECD's data concerning the proportion of major synthetic materials first produced in various countries during the 1930s and early 1940s suggest that the proportion arising from abroad may have been much higher than the figure cited in the text. See *Gaps in Technology: Plastics* (Paris: Organization for Economic Cooperation and Development, 1969). All that we are willing to conclude from these figures is that a substantial proportion of chemical innovations in the United States has been based directly on foreign technology.

24. See Mansfield, *et al., Research and Innovation op. cit.*

ular, Dow (which was a relatively small firm at that time) carried out a relatively large number of developments and innovations in 1930–1950.

Fourth, judging from these and previous studies, it appears that the role of the largest firms in the innovative process varies enormously from industry to industry. In some industries, like steel, the biggest firms have carried out relatively few major innovations; in other industries, like chemicals, the biggest firm seems to have performed very well.[25] Moreover, the results vary, depending on the type of innovation. For example, the biggest chemical firm seems to have played a much more important role in product innovation than in pure process innovation. And the results can vary substantially, depending on whether one looks at innovations, developments, or the basic concepts underlying innovations.[26] This diversity is strikingly at odds with some of the simple generalizations found frequently in the literature.

Fifth, turning to the sources of major chemical innovations, we find that such innovations have been introduced by many firms other than the traditional chemical companies. With the entry of petroleum firms (and paint, rubber, and other firms) into the chemical business, a substantial proportion of both process and product innovations have been carried out by these firms whose primary business is outside chemicals.

Sixth, a significant percentage of major chemical innovations was based on developments carried out by firms other than the innovator. In general, smaller firms are more likely than bigger ones to base an innovation on a development carried out elsewhere. Both in the case of products and processes, many innovations have been based on foreign developments, particularly in West Germany and the United Kingdom. Clearly, technology flows across the Atlantic in both directions, not just one.

Finally, the very considerable limitations of these results should be obvious. Because the sources of specific innovations sometimes are difficult to specify, and the results for various innovations are difficult to weight and combine, as well as for a variety of other reasons, the results are crude. Nonetheless, they should add to our relatively limited knowledge in this area. Without question, studies of this sort are needed to help promote a better understanding of the innovative process.

25. In the electric power industry, a recent study concludes that the largest firms have "not outperformed the other utility classes in relative terms." See B. Smith, "Technological Innovation in Electric Power Generation: 1950–1970," *Land Economics,* November 1974.

26. Also, the results can vary with time (as apparently has occurred in drugs). For example, according to some observers, du Pont now has much less of an edge over its rivals than it used to have. However, it will be some time before this proposition can be tested adequately.

4 THE INNOVATION AND DEVELOPMENT PROCESSES WITHIN THE FIRM

1. Introduction

In the last chapter, we were concerned with the sources of innovation—the types of firms that have been responsible for major innovations. Although data concerning the sources of innovation are valuable for a variety of purposes, they tell us very little about the innovation process within the firm—the process that culminates in the introduction of a new process or product. In this chapter, we are concerned with two aspects of the innovation process. First, we try to obtain a better understanding of why the total costs of innovating are largely R and D costs in the case of some product innovations, but are largely non-R and D costs in the case of other product innovations. The results should shed new light on the role of R and D in the innovation process.[1]

1. For some recent publications concerning the innovation process, see the Organization for Economic Cooperation and Development, *The Conditions for Success in Industrial Innovation* (Paris: 1971); S. Myers and D. Marquis, *Successful Industrial Innovations* (Washington, D.C.: National Science Foundation, 1969); E. Mansfield, J. Rapoport, J. Schnee, S. Wagner, and M. Hamburger, *Research and Innovation in the Modern Corporation* (New York: Norton, 1971); and B. Williams (ed.), *Science and Technology in Economic Growth* (New York: Macmillan, 1973). Also, see National Science Foundation, *Proceedings of a Conference on Technology Transfer and Innovation* (Washington, D.C.: Government Printing Office, 1967); Office of the Director of Defense Research and Engineering, *Project Hindsight* (Washington, D.C.: 1969); IIT Research Institute, *Technology in Retrospect and Critical Events in Science* (Washington, D.C.: NSF, 1968); and Materials Advisory Board, *Report of the Ad Hoc Committee on*

Second, we describe in some detail the anatomy of the development process in polymers, a very important segment of the chemical industry. Development is one of the least understood parts of the innovative process. Klein, Marschak, and Peck and Scherer have provided descriptions of the development process in the aerospace industry, and Comanor, Clymer, and others have discussed the development process in ethical drugs.[2] But for most industries, there is little in the way of such descriptions. In this chapter, the development process for polymers will be divided into eight activities, each of which will be defined and discussed in detail. In the next chapter, we shall build on this account of the development process in polymers, our objective being to analyze, explain, and forecast development costs for particular projects.

2. *The Product-Innovation Process: Structure and Costs*

For present purposes, the innovation process is considered to begin when exploratory work is started by the innovator with the particular new product in mind. Thus, it excludes basic research but includes applied research (if any) directed toward the product. The innovation process ends when the new manufactured product is available (at an acceptable quality level and cost) for sale and delivery. Following the procedures of the Panel on Invention and Innovation of the Department of Commerce, as well as our own previous studies,[3] we classify the work leading up to the first commercial introduction of a new product into the following stages: applied research, preparation of product specifications, prototype or pilot-plant construction, tooling and construction of manufacturing facilities, manufacturing start-up, and marketing start-up. Of course, there is no presumption that these stages do not overlap, or that they must occur in any particular time sequence.

More specifically, the activities included in each stage are as follows: *Applied*

Principles of Research-Engineering Interaction (Washington, D.C.: National Academy of Science, 1966).

2. See B. Klein, "Policy Issues in the Conduct of Military Development Programs," in R. Tybout, ed., *The Economics of Research and Development* (Columbus: Ohio State University Press, 1965); T. Marschak, T. Glennan, and R. Summers, *Strategy for R and D* (New York: Springer-Verlag, 1967); M. Peck and F. Scherer, *The Weapons Acquisition Process* (Cambridge, Mass.: Harvard University Press, 1962); W. Comanor, "Research and Competitive Product Differentiation in the Pharmaceutical Industry in the United States," *Economica* (November 1964); and H. Clymer, "The Changing Costs of Pharmaceutical Innovation," in J. Cooper, ed., *The Economics of Drug Innovation* (Washington, D.C.: American University, 1970).

3. See U.S. Department of Commerce, *Technological Innovation* (Washington, D.C.: Government Printing Office, 1967); and Mansfield *et al.*, *Research and Innovation, op. cit.*

research, as defined by the National Science Foundation (and used here), is "investigation directed toward the discovery of new scientific knowledge where such knowledge is sought with a particular commercial use in mind."[4] An example of the applied research done on products in our sample (described below) is an investigation to find a cheaper method to synthesize a resin previously synthesized by university chemists. *Preparation of project requirements and basic specifications* often consists partly of routine planning and scheduling which involves close coordination with marketing to increase the probability that the proposed product will be a success commercially. In the chemical industry, this stage generally contains some bench-scale work and applications research. The definition of *prototype or pilot-plant design, construction, and testing* seems fairly self-evident. Although the major concern here is the product itself, the work often involves the firm in new technological areas related to manufacture of the product in quantity. Since all of the innovations included in our sample (described below) are chemicals, we are concerned here only with pilot plants, not with prototypes.

The definition of the preparation of detailed manufacturing drawings, tooling, and the design and construction of manufacturing facilities also seems self-evident. While this may lead to minor changes in the product in order to make it a more suitable object for manufacture, the focus is on the problems of manufacturing rather than on the product itself. This stage may represent a transfer of responsibility from the R and D management to the production management and is quite often characterized by considerable investment in physical capital. *Manufacturing start-up* includes training workers, "debugging" the plant, and production before an acceptable quality level is reached. *Marketing start-up* includes activities like market studies, advertising, establishment of a system of distribution, and training the sales force. Marketing start-up cost includes all expenditure of this nature before the first sale and delivery of the product.

Table 4.1 shows, for a sample of thirty eight innovations in the chemical, machinery, and electronics industries, the average and standard deviations of the percent of total innovation costs attributable to each of these six stages. Clearly, as one would expect, there is an enormous amount of variation among innovations in the percent of total costs attributable to each stage. (The standard deviations are eloquent on this score.) The National Science Foundation's definition of research and development accords well with the first three stages listed in Table 4.1. Thus, on the average, R and D accounted for about 40 percent of the total innovation cost for the sample of chemical innovations, and for about 50 percent of the total innovation costs for the sample of machinery innovations and the sample of electronics innovations. In each industry, there is a great deal of variation

4. National Science Foundation, *Methodology of Statistics on Research and Development* (Washington, D.C.: 1959).

TABLE **4.1**

Percent of Total Cost of Innovation Arising in Each Stage of Innovative Activity

STAGE	CHEMICALS	MACHINERY	ELECTRONICS	WHOLE SAMPLE
		(percent)		
Applied research				
Mean	17	3	4	10
Standard deviation	17	5	6	
Specifications				
Mean	13	4	3	8
Standard deviation	17	6	3	
Prototype or pilot plant				
Mean	13	41	44	29
Standard deviation	8	17	17	
Tooling and manufacturing facilities				
Mean	41	37	30	37
Standard deviation	29	10	18	
Manufacturing start-up				
Mean	8	4	14	9
Standard deviation	8	6	9	
Marketing start-up				
Mean	7	11	6	8
Standard deviation	12	12	5	

Source: E. Mansfield, J. Rapoport, J. Schnee, S. Wagner, and M. Hamburger, *Research and Innovation in the Modern Corporation* (New York: Norton, 1971).

among innovations in the percent of total innovation costs going for R and D.[5]

To help shed new light on the factors associated with the large differences among innovations in the percent of innovation costs devoted to R and D, detailed data were obtained concerning seventeen chemical innovations completed in the 1960s. Admittedly, this is a small sample, but because a great deal of detailed information had to be collected concerning each innovation, this was as big a sample as the available resources would support. These innovations were introduced by five chemical firms, and were primarily polymers for industrial use (such as synthetic elastomers and fibers) and a few organic chemicals (such as starches). The innovations included had all been commercialized (or very nearly so) at the time we obtained the data, and each had been carried out entirely by

5. Neither here nor in subsequent sections are we assuming that the observed distribution of costs for each innovation was the distribution planned by the firm at the outset of the project. Obviously, the actual distribution is likely to differ substantially from the planned distribution.

one firm. Data on the distribution of total cost (among the six stages, defined above) for each innovation were obtained through personal interviews with managers who actually participated in the work on the project. Since firms' accounting records often did not yield data corresponding exactly to the breakdown into stages we used, accounting data were supplemented by estimates of the managers. Other relevant information about the innovations was obtained during the interviews, by a later mail questionnaire, and by follow-up conversations and interviews.

The following statistics provide some idea of the characteristics of the innovations included in the sample: The average total innovation cost was $2,252,000 (the standard deviation being $2,318,000); the average annual expected revenue from the product during the first three years of its life was $5,941,000 (the standard deviation being $13,017,000); and the average number of years of R and D experience of the innovator in the relevant product area was 8.8 (the standard deviation being 9.8). Further, in 35 percent of the cases, the innovation was a response to a similar product by a competitor (and in 65 percent, it was not); in 29 percent of the cases, it was necessary to construct a new pilot plant (and in 71 percent, it was not); in 59 percent of the cases, all or most of the equipment required to produce the product was already available (and in 41 percent, it was not); and in 53 percent of the cases, the firm's workers had considerable experience with the production of this kind of product (and in 47 percent, they did not).

3. Factors Influencing the Percent of Innovation Costs Going for R and D

In our sample of seventeen chemical innovations, R and D costs accounted for 39 percent, on the average, of total innovation costs. But, as in Table 4.1, there is an enormous amount of variation about this average. Specifically, the percentage of total innovation cost attributable to R and D ranged from 7 percent to 71 percent. Clearly, some innovations were accomplished with a relatively small expenditure on R and D, whereas, for other innovations, the R and D costs were the lion's share of the total innovation costs. What factors are responsible for these great differences? Why is it that R and D bulks large in the costs of some innovations, but is relatively insignificant in others? To help answer these questions, we see how well a number of hypotheses—some suggested by economic theory, some suggested by conversations with leading chemical engineers—can explain the observed variation. Some of these hypotheses are extremely tentative, since the available theory in this area is so weak. We regard them as interesting hypotheses to test, even though we do not feel that, on a priori grounds, all of them are equally convincing.

The first hypothesis is that bigger, more important innovations—ones that constitute major new products that enjoy large markets—require, on the average, a larger proportion of their costs to go for R and D. It is easy to point to exceptions, but it is frequently argued that this hypothesis holds on the average. Whereas relatively small improvements can be made without much R and D, it is argued that, on the average, this is not the case for more important innovations. To measure a product's importance, we use its expected annual sales in the first three years of life. This measure is far from ideal, but it should be good enough for present purposes.[6]

The second hypothesis is that, holding other factors constant, larger firms may devote a larger percentage of total innovation costs to R and D than smaller firms. Although there is bound to be great variation in individual cases, this hypothesis seems reasonable on the average. R and D, because of its uncertain nature, may produce results in areas not directly related to the specific problem at hand. A large firm may feel that it is more likely to be able to utilize such unpredictable spillovers, and thus may be willing to spend more of the total costs on R and D than a smaller firm. For another thing, the available evidence suggests that it often costs a large firm more than a small one to develop the same product.[7] If this is the case, and if the difference between large and small firms in non–R and D costs is smaller, one would expect the percent of total costs attributable to R and D to be higher for the large firms than the small ones.

The third hypothesis is that the extent of the innovator's R and D experience in the relevant technological area is likely to affect the percent of the innovation's costs that go for R and D. Economists have devoted considerable attention to the phenomenon of "learning by doing" in production, but they have been slow to recognize the importance of "learning by doing" in R and D. This is unfortunate, because organizational learning is important in R and D too. In R and D, as in production, the more experience that a firm has in the relevant technological area, the more efficient it is likely to be. Since there is probably much less relationship between a firm's R and D experience in the relevant area and the amount of learning in the areas of tooling and construction of plant, manufacturing start-

6. Expected sales, rather than actual sales, were used because some of the innovations were only entering the market at the time when the data were collected. Of course, the choice of three years is arbitrary. We could just as well have used five years or some other period. But three years seems reasonable, at least for these innovations. However, we recognize that, for very novel or far-reaching innovations, this period might underestimate the importance of an innovation.

7. See E. Mansfield, *Industrial Research and Technological Innovation* (New York: Norton, for the Cowles Foundation for Research in Economics at Yale University, 1968); A. Cooper, "R and D Is More Efficient in Small Companies," *Harvard Business Review* (June 1964); J. Schmookler, Testimony before the Senate Judiciary Committee (Washington, D.C., 1965); and E. Mansfield *et al., Research and Innovation, op. cit.*

up, and marketing start-up, one might expect the percent of total costs attributable to R and D to be lower for firms with more R and D experience in the relevant area.

Finally, two other variables that may be relevant are the size of the total innovation costs and whether or not a new pilot plant must be built. As the scale of a project increases, the non–R and D costs, particularly for tooling and construction of manufacturing facilities, may not tend to increase in proportion with the R and D costs. Thus, the size of the total innovation costs may have a systematic effect on the percent of the total innovation costs attributable to R and D. Similarly, whether or not a new pilot plant must be built may also affect the percent of an innovation's costs that go for R and D. However, it is difficult to predict the direction of this effect. On the one hand, one might expect an innovation where a new pilot plant must be built to have relatively more spent on R and D simply because of the necessity of building a pilot plant. On the other hand, innovations requiring a new pilot plant are likely to be much more ambitious projects where the non–R and D costs are much higher too. Consequently, it is hard to tell whether the net effect of this variable is to reduce or increase the percent of total costs going for R and D.

4. Econometric Results

To see how well these hypotheses can explain the observed variation, we assume that

$$D_i = \alpha_0 + \alpha_1 R_i + \alpha_2 S_i + \alpha_3 E_i + \alpha_4 T_i + \alpha_5 N_i + z_i, \tag{4.1}$$

where D_i is the percent of the ith innovation's total costs spent on R and D, R_i is the expected annual sales of the ith innovation (over the first three years of its product life), S_i is the annual sales of the firm that was the innovator in the case of the ith innovation, E_i is the number of years of R and D experience in the relevant technological area of the innovator in the case of the ith innovation, T_i is the total cost of the ith innovation (the sum of the six stages in section 2), N_i is a dummy variable that equals 1 if a new pilot plant is built in the case of the ith innovation and zero otherwise, and z_i is a random error term. S_i is measured in millions of dollars, while R_i and T_i are measured in thousands of dollars.

Using the data concerning our sample of seventeen innovations, we computed least-squares estimates of the αs. The results are:

$$\begin{array}{l} D_i = 37.4 + 0.00076\, R_i + 0.019\, S_i - 0.59\, E_i - 0.0051\, T_i - 11.9\, N_i \\ \qquad\qquad\quad (3.2) \qquad\qquad (3.1) \qquad (1.7) \qquad (2.8) \qquad\quad (1.9) \end{array} \tag{4.2}$$

The t ratios are given in parentheses below each regression coefficient. This regression fits the data reasonably well, $\bar{R}^2$—after adjustment for degrees of freedom—being 0.55. In other words, this regression explains about 55 percent of the variation among these innovations in the percent of total innovation costs devoted to R and D. It is not surprising, of course, that a large proportion (45 percent) of the variation remains unexplained. In view of the enormous heterogeneity of product innovations in the chemical industry, and the many other factors that influence D_i, we would not expect a very high value of $\bar{R}^2$.

Each of the regression coefficients in equation (4.2) has the expected sign (in those cases where it is stipulated in advance). In accord with the first hypothesis, the size of a new product's expected annual revenue (R_i) seems to have a major effect on D_i. Specifically, if R_i increases by $13 million (one standard deviation), the percent of total innovation costs devoted to R and D increases, on the average, by about 10 percentage points. Also, the effect of the size of the innovating firm seems to be in accord with the second hypothesis. Judging from equation (4.2), a $250-million increase in the annual sales of the innovator will result in an increase of about 5 percentage points, on the average, in the percent of total innovation costs devoted to R and D. Since the firms in our sample ranged from $100 million to over $2 billion in annual sales, it is evident that variation in firm size within the range of sizes represented in our sample seems to be associated with substantial differences in the percent of total innovation costs going for R and D.

In accord with the third hypothesis, the extent of the innovating firm's R and D experience in the relevant technological area also seems to have a significant effect. But unlike the other variables, it is significant only at the 10-percent, not the 5-percent, level. Judging from equation (4.2), ten years of additional experience (one standard deviation) reduce the percent of total innovation costs devoted to R and D by about 6 percentage points, on the average. Also, the results suggest that increases in total innovation costs are associated with reductions in the percent of such costs going to R and D, and that innovations where a new pilot plant is built tend to spend a lower percentage of total costs on R and D. Apparently, in cases where a new pilot plant is built, the added non–R and D costs tend to be bigger, percentagewise, than the amount that the pilot plant adds to the R and D costs.

5. *Implications*

Economists sometimes have assumed that research and development expenditures could be treated as synonymous—or nearly so—with the costs of product innovation. In previous studies, it has been shown that this assumption is a poor one, since R and D expenditures account, on the average, for less than half of the amount spent on a product innovation (in the cases we studied). However, there is an enormous amount of variation among innovations in the percent of total innovation costs that go for R and D. The results of the previous section suggest that a substantial proportion of this variation can be explained by the importance of the innovation (as measured by sales volume), the amount of R and D experience the innovator has had in the relevant technological area, the size of the innovating firm, the total innovation costs, and whether or not a new pilot plant must be built.

In view of the smallness of the sample (composed entirely of chemical innovations) and the roughness of the data, these results should be viewed as highly tentative. However, if confirmed by future research, they seem to have several interesting implications. First, if R and D plays a more important role in more major innovations than in less important ones, this means that, if innovations are unweighted by their importance (as so frequently is the case), data concerning the percentage of innovations based to any significant degree on R and D are likely to contain a serious downward bias. Yet data of this sort have often been used in important policy discussions (in this country and abroad) concerning the effects of R and D on technological change.

Second, the results should be of interest to students of industrial organization. If "learning by doing" is as important in R and D as it seems, this fact has implications for competitive relationships among firms, and for barriers to entry. Clearly, firms with a headstart in a particular area of technology can gain a substantial advantage over Johnny-come-latelys. Further, if it is true that large firms tend to devote a higher percentage of total innovation costs to R and D, the official figures on R and D expenditures may exaggerate the extent to which total expenditures on innovation are concentrated in big firms. This, of course, is interesting from the point of view of public policy toward the large firm.

Finally, the results indicate (for the types of innovations considered here) the extent to which a prospective innovator must expect to supplement its R and D funds in order to bring a new product to market, and some factors associated with the size of this supplement. On the average, this supplement was two or three times as great as the R and D costs, depending on the industry. And expressed as a percentage of the R and D costs, this supplement was smaller for more important innovations than for less important ones, and for big firms than small ones. To the potential innovator, the size of these supplementary costs is a matter of considerable importance.

6. The Development Process in Polymers

In previous sections of this chapter, we have been concerned with the entire innovative process, and the ways in which the costs of innovation are distributed among various activities. In the rest of the chapter, we narrow our focus to one part of the innovative process—development. Development is sometimes defined as "the technical activity concerned with nonroutine problems which are encountered in transforming research findings or other general scientific knowledge into products or processes."[8] While this is an accurate definition, it does not tell us a great deal. The purpose of the rest of this chapter is to describe in detail the anatomy of the development process in polymers.[9] This analysis will be used in the following chapter, where we construct an econometric model to explain and forecast development costs in polymers.

One major reason for choosing polymers is the importance and phenomenal growth of the polymer industry. The production of polymers (including rubbers, plastics, and synthetic fibers) has grown more rapidly in this century than any other sector of the chemical industry. Worldwide production of plastics, just one branch of the polymer family, expanded from a total of 2.9 million tons in 1955 to 16.3 million tons in 1966, an average rate of growth of 17 percent. In 1969, $482 million was devoted to applied research and development in plastic materials and synthetic resins, rubbers, and fibers in the United States, of which only $15 million (3.1 percent) was funded by the federal government. Many of the innovations cited in the previous chapter were polymers.

Development is comprised, of course, of a network of related activities moving forward simultaneously. In separating and classifying these activities, we want to make the breakdown specific enough for meaningful boundaries to be placed on each activity or phase, but general enough so that application is possible to a broad range of development projects carried out in different firms. This is by no means a simple task. It was only after lengthy discussions with many R and D managers, chemical-engineering consultants, and university professors that a classification scheme, which seemed acceptable to practically all, was formulated. The process for industrial polymer development was finally divided into the following eight activities:

1. preliminary engineering studies;
2. toxicity studies;
3. applications research;
4. marketing research;
5. patent and legal studies;

8. National Science Foundation, *op. cit.*, p. 124.

9. Polymerization is the chemical process in which small molecules (monomers) are linked together to form one large molecule. Depending on the length, detailed shape, and chemical structure of the molecule, an astonishing number of desirable properties can be conferred upon these materials. They an be produced in the form of solid plastics fibers, foams, films, coatings, and adhesives. For a more detailed discussion of polymers, see H. R. Simonds, *A Concise Guide to Plastics*, 2nd ed. (New York: Reinhold, 1963), or H. R. Simonds and J. M. Church, eds., *The Encyclopedia of Basic Materials for Plastics* (New York: Reinhold, 1967).

6. bench-scale work;

7. pilot-plant work;

8. market development.

As will become evident in the next chapter, these activities are by no means equal either in terms of the cost or the time devoted to each. Nor is a temporal sequence of these activities to be inferred from the above ordering. The breakdown is meant simply as a concise, although somewhat arbitrary, means of dividing the development process into more meaningful constituent activities. The problem of defining the respective boundaries of each of these activities was a major one which could be handled only by carefully detailing the respective definitions to each participating firm.[10] For most firms these activities had distinct accounting identities, but for some this was not the case and costs had to be estimated. In the next eight sections of this chapter, we define each of these activities and discuss their nature and significance. As we shall see in the next chapter, this classification scheme can be applied to development projects aimed at both new products and new processes.

7. *Preliminary Engineering Studies*

On the basis of data derived primarily from the applied-research effort, preliminary engineering studies generally attempt to examine and evaluate the development project's economic and technological feasibility before the final commitment is made to advance the project beyond the applied-research phase to the more costly phases of development. The polymerization process is analyzed, and an estimate of capital costs and operating costs may be made.[11] A feature often distinguishing this activity from the applied-research stage which precedes it and the bench-scale and/or pilot-plant activities which follow is that much of the work is likely to be carried out on paper or by a computer rather than in the experimental equipment of the laboratory or plant.

In general, the purpose of the preliminary engineering studies is just as the title implies—to provide a preliminary feasibility study to determine whether a full-scale development project should be undertaken. The question which is up-

10. After lengthy discussion with participating (as well as nonparticipating) firms in this study, but before any hard data had been collected, a detailed summary of the purpose of the study and the definition which had been constructed for each activity was circulated to the various firms.

11. For some relevant discussion, see R. N. Shreve, *Chemical Process Industries*, 3rd ed. (New York: McGraw-Hill, 1967).

permost in the minds of the development managers is (or at least should be): Is this project economically and technologically feasible? The extent of the effort allocated to the preliminary engineering studies will obviously depend on the marginal cost of such efforts and the marginal benefits derived therefrom. The factors which might be expected to influence the costs of this activity relate to the general uncertainty underlying the project, which in turn relates to the scope of the project, the expertise of the firm in this area, and the facilities which the firm possesses with respect to the technology necessary for the particular project. The adage that a firm should "make its mistakes on a small scale and its profits on a large scale" [12] is relevant, of course, in undertaking these feasibility studies.

In many projects, particularly those which involve only a minor modification or improvement, little or no expenditure is necessary for preliminary engineering studies. The firm is often so familiar with the technology and with the product that development will involve only bench-scale and/or pilot-plant work. However, there are also many projects in which these preliminary engineering studies are an important and costly component of the development process. In terms of NSF definitions, some preliminary engineering studies may be regarded as applied research, not development. But since many firms in our sample regard prelimary engineering studies as an integral part of the developmental process, it seemed sensible to include them. Of course, if one wants to exclude them, it is simple to modify our results accordingly.[13]

8. *Toxicity Studies*

This activity was defined to include studies to obtain preliminary toxicity information on the material, using tests to investigate acute oral toxicity, skin and eye irritation, vapor exposure, and others. This activity was included primarily upon the advice of a number of chemical-engineering consultants who had had previous experience with the development of polymers for either clothing or food-packaging purposes. As it happened, only two projects in our sample involved any expenditures for toxicity studies, and in these cases the expenditure was minor.

12. This quote is attributed to L. Baekeland, the originator of Bakelite, a thermosetting plastic.

13. Actually, these preliminary engineering studies do not tend to be very costly in the sample of fifty-six projects discussed in Chapter 5. There were nineteen which involved no preliminary engineering costs, another eight which involved costs of $2,000 or less, and another eleven which had costs of $5,000 or less. Thus, it does not make much difference whether they are included or not.

9. Applications Research

This activity was defined to include studies in the laboratory or plant to determine the fundamental mechanical, thermal, electrical, and other properties of a new or improved product with careful attention being directed to the ultimate fabrication and uses of the polymer or plastic material. More specifically, it consists of determining the physical and chemical properties of the product generated from the bench-scale or pilot-plant apparatus as a function of composition of matter, and/or formulating the material in order to optimize its behavior in various applications. The primary purposes of this phase of the development process are fourfold:

1. to thoroughly examine the properties of the polymer or family of polymers in the case of an entirely new polymer;
2. to test the application of the product in desired or existing uses;
3. to discover possible new uses for the product;
4. to examine the ease of fabrication of the product as well as the possibility of different modes of fabrication.

The extent of the work performed in this phase will depend on the degree of departure of the polymer from the usual product lines of the firm and also on the firm's experience with particular marketing applications[14] and fabricating procedures. This activity can best be depicted as a "spiral" process by which the new polymer or product is put through successively more detailed rounds of synthesis, characterization, and evaluation for utility. Since the real test of a material comes with actual service, the ultimate goal of this activity is to test and prepare a product to meet the characteristics desired by the ultimate consumer.[15] Of course, applications research generally must be carried out in projects aimed at new processes, as well as at new products. In the case of new processes, the

14. Each "marketing application" can be thought of as one member of the five-digit Standard Industrial Classification (SIC) code.

15. One might argue that, for the case in which a product is being developed with no specific application in mind, the work in the application-research phase might be more typical of research rather than development work. However, the counterargument is that once a project is formally proposed and funds are committed to carry this project to its commercial introduction, management has a concrete notion of the possible markets in which this product may be profitable, even if the product is not being developed for a specific need. Often there are one or more key properties which arouse interest in the product, and these properties are exploited. Also, this activity may be continued after the innovation is commercialized, as the firm explores the possibility of new or improved applications, but such costs are not included here since they are not required for the product's original development.

characteristics of the product must be retested to insure that they have not been altered by variations in the process.

10. Marketing Research

This activity was defined to include studies and surveys designed to estimate the probable extent and location of the market, and to determine the acceptable grades and specifications for a product in relation to final selling price and compared with alternative or competitive products. (Frequently, no marketing research is carried out for new processes.) Functionally, this activity consists of studies of competition, prices, and channels of distribution. The primary purpose of these studies is to answer the question: Given the technical success of the product, the substitutes for the product, and the competition, can this product be marketed at an acceptable profit to the firm?[16] While this type of work is explicitly excluded from the definition of development put forth by the National Science Foundation, it was decided that, since many firms seem to regard marketing research as an integral part of the development process, we would include it. However, this makes little difference, since, as we shall see, the costs of this activity do not, on the average, add very significantly to the total cost of development in our sample of projects. Moreover, if one wants to exclude this activity, it is easy to modify our results accordingly.[17]

11. Patent and Legal Studies

This activity was defined as any work performed by the patent department or outside law firms to determine the position of the product or process with respect to its patentability. Prior to obtaining the results discussed below, we heard from many people that patent and legal costs were important parts of the costs of development.[18] While there is no doubt that patent and legal considerations play an

16. Of course, some of these "givens" must be determined by the marketing-research studies.

17. There is, however, a noteworthy problem concerning the measurement of the cost of the marketing-research activity for individual projects. While there is most likely some error involved in the measurement of the cost of the other phases of development due to accounting idiosyncracies, failure to report costs under the appropriate accounts, etc., this is much more of a problem in attempting to estimate the marketing-research cost by individual project. Most marketing-research staffs are responsible for many separate projects, and the costs of the staff are not allocated to individual projects. Hence, in most cases these costs had to be estimated and were generally recorded only if they were substantial.

18. For a discussion of the legal costs of patenting as they relate to R and D expenditures, see R. L. Sandor, "Some Empirical Findings on the Legal Costs of Patenting," *Journal of Business,* XLV/3 (July 1972).

important role in the development process, the organizational milieu of most firms makes it very difficult to attribute patent and legal costs to individual projects unless such costs are substantial. For this reason, even though data concerning such costs were collected for this activity, they may not be very accurate. If one wants to exclude this activity, our results can be modified accordingly. The NSF definition of development excludes such work.

12. Bench-Scale Work

For the purposes of this study, bench scale was defined as a setup of equipment to carry out the key chemical reactions and operations involved in the manufacture of the product. The primary emphasis of the bench-scale work is on solving technical, as opposed to marketing, problems. The work is generally carried out in relatively inexpensive equipment (often stock glassware items), and the experiments may be performed on a scale ranging from a small test tube to a multi-liter reactor. A distinguishing feature of bench-scale work is that, although it is small-scale and relatively inexpensive, an attempt is made to come as close as possible to duplicating the operation of the ultimate plant.

There are obvious difficulties with performing various units of experimentation with glassware when ultimately the process will take place in large reactors. However, the main task of the scientists on the bench-scale level is to reduce the uncertainty which might preclude an advance to larger-scale work, while at the same time minimizing development costs. Since there is often a vast difference in the scale of operation between bench-scale and pilot-plant work, it is preferable to perform as much as possible of the technical problem solving on a bench-scale level so that problems of plant design, scale-up, and production for customer evaluation may be concentrated on at the pilot-plant level. Of course, bench-scale work is carried out in research as well as in development, but we are concerned here with development only.

13. Pilot-Plant Work

A pilot plant may be defined as a "scale model—built of engineering materials and operated for a relatively short period of time—of a larger chemical plant."[19] The pilot-plant operation, which is often the most lengthy and expensive phase of the development process, is generally carried out in apparatus which resembles production equipment and which is designed to simulate the projected conditions of the commercial process.

19. D. G. Jordan, *Chemical Pilot Plant Practice* (New York: Interscience, 1955), p. 5.

Two primary factors distinguishing pilot-plant from bench-scale work are *scale* and *expense of operation*. Generally, as many problems as possible are considered and solved at the bench scale, leaving those which can't be solved satisfactorily for the pilot plant. The pilot plant may be operated for any one of a number of reasons, and these reasons often are difficult to distinguish. While the physical specifications of the pilot plant may be the same in each case, three different modes of operation for the pilot plant may be distinguished for our purposes:

1. *technical*—The unit is built or operated primarily to produce technical information. Engineering data may be measured for a new situation which is not covered in the literature or by the firm's experience. In this type of work, little attention is paid to production, and most attention is centered on quantitative measurement.

2. *scale-up*—The completely integrated plant, a scale model of the proposed unit, is built and operated continuously (if possible and/or desirable). Every effort is made to improve yields and conversions, to test materials of construction, to evaluate by-products, and to produce quantities of the product for final evaluation (i.e., market-development purposes).

3. *process and product evaluation*—In this instance, the main intent is simply to determine that the product resulting from this process satisfies proposed specifications.

While there are many instances in which the pilot plant is operated to satisfy just one of the above requirements, it is often the case, especially in large development projects, that each of the above purposes is served at some time during the life of the project. In fact, pilot-plant work may be thought of as running the gamut from the simple operation of existing pilot-plant or plant facilities for a short period of time to determine simply whether the process works, to the construction of a multimillion-dollar semiworks plant to investigate thoroughly all facets of the plant operation.[20]

Prior to any major commitment to pilot-plant work, the firm will have a specific set of goals in mind. The pilot plant is too expensive for undesigned experimentation. The final production unit can often be visualized from the preliminary

20. The cost of pilot-plant work for the projects studied in Chapter 5 ranged from $0 to $3,990,000. Of the fifty-six projects, forty-six incurred positive costs in the pilot-plant phase, but only four necessitated the construction of a pilot plant or semiworks. The remainder of the projects in which any pilot-plant expense was incurred involved work in existing pilot-plant or plant facilities.

According to many chemical engineers, pilot-plant work has been reduced in recent years by the development of new analytical devices. This should be borne in mind in interpreting our results in Chapter 5.

engineering and bench-scale studies, and the types of information required for final plant design listed. The function of the pilot plant, which will depend on the information requirements and objectives, will be set accordingly. These specific objectives will naturally depend on the particular situation. In general, if a new product or process is being developed, the main concern is with the collection of information for the design of the commercial plant and for economic evaluation of the product and process. On the other hand, if proposed changes in an existing product or process are being investigated, the usual objectives are to test the minor variations which are involved in the product or process and to ensure that the modification emerges as projected when the process is scaled up.

14. Market Development

The market-development activity is defined as the *promotional* effort between applied research and regular sales programs by which existing or potential markets for new products and entirely new fields of utility for old products are developed and tested for profitability. More specifically, market development embraces all the steps involved in taking a new or improved product to potential customers, evaluating its performance in the latter's plants, obtaining their reactions, and attempting to interest them in the innovation. On an organization chart, this is one of the most difficult activities to locate, for market-development work may be performed by people in the sales department, in the technical-service department, or even on the pilot-plant staff. Yet market development is one of the most significant phases in the commercial development of a new chemical product.

The importance of coordinating the marketing and R and D functions cannot be overemphasized. It lies at the heart of successful commercial development. As pointed out in Chapter 2, there has sometimes been a tendency to neglect the marketing aspects of R and D while concentrating on the technological aspects. This may be one reason for the poor track record of some firms in developing commercially successful products.[21] Recently, firms seem to have become more acutely aware of the need for close cooperation between the R and D and marketing departments.

The personnel involved in the market-development effort must consider the prospective client carefully. This may first involve taking samples directly from the laboratory to the prospective customer, displaying the attractive properties of the material, and ferreting out the customer's reaction to the product. During the development process, the people involved in market development are continually trying to gain new customers in the proposed markets and even trying to find new

21. See Chapter 2 above, as well as Chapters 3 and 10, of Mansfield *et al.*, *Research and Innovation, op. cit.*, and Booz, Allen, and Hamilton, *Management of New Products* (New York: 1960).

markets in which the product may gain a substantial foothold. Of course, their success depends on how flexible the people in other phases of the development program are in adapting the material or product to the various customers' needs.

Market development is excluded from the NSF definition of development. If one wants to exclude this phase, our results can easily be modified accordingly. (In the model cited in the next chapter, a separate equation is estimated for market-development costs.) Note too that market-development and marketing-research costs are part of marketing start-up costs in section 2.

15. Summary and Conclusions

Despite its importance, relatively little is known about the innovation process within the firm. Even the costs involved in this process are not well understood. Although there is considerable evidence that R and D typically accounts for no more than about half of total innovation costs, little is known about the reasons why R and D is a very large proportion of innovation costs in some cases, but a very small proportion in others. To throw some light on this question, we obtained detailed data concerning the composition of innovation costs for seventeen new chemical products. Then we constructed and tested a simple econometric model to explain the differences among the innovations in the proportion of total innovation cost going for R and D.

This model was based on several tentative hypotheses. The first hypothesis is that bigger, more important innovations require (on the average) a larger proportion of their costs to go for R and D. The second hypothesis is that, all other things equal, larger firms tend to devote a larger proportion of total innovation cost to R and D than smaller firms. The third hypothesis is that the extent of the innovator's R and D experience in the relevant technological area is likely to affect the proportion of an innovation's costs that go for R and D. Also, the scale of the project may affect the proportion of an innovation's costs for R and D.

This model can explain about 55 percent of the observed variation among innovations in the proportion of total innovation costs going for R and D. In view of the smallness of the sample and the roughness of the data, these results should be viewed as highly tentative. But if confirmed by future research, they seem to have several interesting implications. For one thing, they seem to imply that unweighted data concerning the percentage of innovations based to any significant degree on R and D are likely to contain a serious downward bias. This is important because such data have been used in some major policy discussions. Another implication of the results is that "learning by doing" seems to be an important factor in R and D, as well as in production. This is important for competitive relations among firms and for barriers to entry, as well as for intrafirm decision making.

Development is certainly one of the most important parts of the innovation process. To promote a better understanding of the development process, we constructed a classification scheme for the activities carried out in the course of polymer development. Polymers are, of course, a key part of the chemical industry, and such a classification scheme should provide a framework for the analysis of what goes on in polymer-development projects. Based on discussions with many chemists, chemical engineers, and managers in chemical firms, engineering firms, and universities, we defined eight activities that comprise the development process in polymers—(1) preliminary engineering studies, (2) toxicity studies, (3) applications research, (4) marketing research, (5) patent and legal studies, (6) bench-scale work, (7) pilot-plant work, and (8) market development.

There is no presumption that these activities occur temporally in the order listed above. On the contrary, as we shall see in the next chapter, the temporal ordering can vary considerably. The definition of development used here is broader than that of the National Science Foundation, but in accord with some firms' definitions. It is easy, however, to omit those phases that NSF does not consider as development, if this seems desirable.[22] In the next chapter, we shall use this classification scheme to analyze the cost of polymer-development projects.[23]

22. We are *not* suggesting that these other phases should be considered part of development by NSF. We include all eight activities, since this allows one to define development more broadly if this seems desirable. Clearly, for many purposes, all eight activities include too much. But, as emphasized in the text, *one can omit those activities that are irrelevant, and tailor the definition of development to the problem at hand.*

23. For a valuable description and analysis of the R and D process in the chemical industry, see A. Baines, F. Bradbury, and C. Suckling, *Research in the Chemical Industry* (London: Elsevier, 1969).

5 AN ECONOMETRIC ANALYSIS OF DEVELOPMENT COSTS

1. Introduction

In the previous chapter, we divided the development process for polymers into eight activities or phases. In this chapter, we look at the distribution of cost and time among these activities in a sample of fifty-six development projects carried out in six firms from 1954 to the early 1970s. This sample is perhaps the largest and richest data base of this sort to have ever been assembled for an econometric study of this kind. Using this data base, we can describe more accurately than has heretofore been possible the amount spent on specific activities at various points in time in various development projects. Of course, the results pertain only to polymer development projects. The development process varies so much from industry to industry that one cannot extrapolate the results in any simple way to other industries. But our findings should illuminate important economic aspects of the development process in the chemical industry.

Besides describing various features of the development process, this chapter summarizes an econometric study of the factors determining the development costs for a particular project. The overall contours of this model are discussed, as well as its predictive ability. Because of limitations of space, only part of the model is considered in detail; but this should be sufficient for present purposes.

2. Overview of the Polymer Development Process: The History of a Project

To make it easier to understand each of the activities described in the previous chapter, and the way that each activity fits into the complicated, interrelated pattern of development, we begin this chapter by describing the history of a particular development project carried out by one of the firms in our sample. The project, which resulted in a totally new product for the firm, cost $761,000 and took about sixty-four months to complete. Table 5.1 shows a breakdown of the project's costs.

The project, which we call project X, began when exploratory work (basic and applied research) yielded a polymer which seemed to have attractive properties.[1] The firm decided that, even though a specific need for the product was not evident, the project should be advanced to the development stage. Among the major problems encountered by the firm during the early stages of development were the following: First, the firm had to test various catalysts to find an appropriate one. Second, there were a number of different "mediums" or systems in which the polymerization could take place, and it was important to determine the optimal one. Third, the firm had to find the optimal temperature and pressure, as well as a satisfactory way to achieve reproducibility.

During the first two years of the development project, the bench-scale work, which cost over $180,000, was the principal activity. The purpose of the bench-scale work was to solve as many of the technical problems as possible. This included detailed study of the problems mentioned above as well as the examination of many other more familiar aspects of the product and process. The emphasis of this work was on control of the molecular weight of the product through control of the polymerization variable. Approximately five polymerization systems were explored and tested for this purpose.

The movement to the pilot plant in March 1967, prior to the end of the first stage of bench-scale work, was prompted by the need to make samples of the product. While the firm was concentrating on obtaining data for the scale-up of the polymerization process, it also wanted to produce material for applications research and further bench-scale work. Hence, the initial pilot plant was essentially a scale-up operation designed both to investigate process variables and to produce some material for further testing.

The movement into applications research in September 1967, after the project had been ongoing for two years, was intended primarily to determine where the polymer could best be utilized. Since it was a brand-new product, tests were necessary in order to determine its most profitable uses. While a number of uses were being explored within this activity, the concentration was on one specific application.

As shown in Table 5.1, various activities—bench-scale work, pilot-plant work, marketing research, applications research, and market development—were carried out in 1968 and 1969. As time went on, fewer and fewer variables were studied in each activity, but each variable was tested more and more completely. The work in each activity was affected by the feedback from work in the other activities, as well as by the feedback from the customer.

1. The cost of this research is not included in Table 5.1. It is not part of development, as defined here (and by NSF).

TABLE 5.1

Data Regarding Cost and Time for Project X

PHASE	TIME	DIRECT LABOR COST ($)	OVERHEAD ($)	MATERIAL ($)	CAPITAL ($)	TOTAL ($)
Preliminary engineering studies	None	—	—	—	—	None
Toxicity studies	None	—	—	—	—	None
Applications research	9/67–12/67	9,000	7,000	—	—	16,000
	5/68–11/68	20,000	16,000	—	—	36,000
	7/70–12/70	5,000	4,000	—	—	9,000
Marketing research	5/69– 9/69	8,000	2,000	—	—	10,000
Patent and legal aspects	5/67– 9/67	4,000	1,000	—	—	5,000
Bench scale	8/65– 8/67	100,000	80,000	2,000	—	182,000
	8/67– 6/68	30,000	25,000	1,000	—	56,000
	5/69– 9/69	7,000	6,000	1,000	—	14,000
Pilot plant	3/67– 4/67	18,000	16,000	1,000	—	35,000
	10/67–11/68	110,000	90,000	5,000	—	205,000
	1/70–12/70	85,000	63,000	5,000	—	153,000
Market development	5/68–12/70	32,000	8,000	—	—	40,000
Total	*8/65–12/70*	*428,000*	*318,000*	*15,000*	—	*761,000*

As the project progressed, the product moved closer and closer to the customer. While the initial feedback from the customer was very weak, it was crucial. If one were to examine the project in more detail, one would become acutely aware of the customer-oriented nature of the work. Although the effort was concerned with solving various technical difficulties, both in the process and product, the main concern was with the customer and with the feedback from the customer.

In the last year of the project, the emphasis was on pilot-plant work (in the last thirteen months, 36 percent of the pilot-plant work in terms of cost was completed) and market-development work. During this period the main function of the pilot plant was to produce a large amount of material for the market-development effort, as well as to examine further the process variables. While the market-development effort was aimed at a number of customers (ten), the main concentration was on one customer. It is obvious from the amount of time and effort devoted to this customer that a great deal of testing, feedback, and further testing was necessary.[2] The product was finally ready for commercialization sixty-four months after the commencement date.[3]

2. Much of this testing was performed within the customer's plant. This generally required not only a large amount of material but also a highly intensive labor effort on the part of the developer.

3. It should be noted that after commercialization, development work was continued on this product in order to determine new and better uses. This is typical of many prod-

3. Relative Costs of Each Activity

Having described the activities comprising the development process in Chapter 4, we now turn to an investigation of the proportion of total development costs resulting from each activity. Table 5.2 presents the average (unweighted) proportion of total development costs accounted for by each of the eight activities for the fifty-six development projects in our sample, together with the standard deviations of these proportions.[4] (All of these development projects were technically completed and commercialized, although not always commercially successful.) In column 1 of this table, these proportions are presented for the total sample, while in column 2 they are presented for only those projects in which the observation was nonzero (i.e., the activity was performed).[5] Finally, in column 3 is presented the proportion of nonzero observations within each category. This last statistic indicates the relative frequency with which each activity was performed in the development process in our sample.

Any a priori notions about the importance of the pilot-plant work in the development process are confirmed in this table. For not only was pilot-plant work required in the highest proportion of projects (.82), but it also accounted, on the average, for the highest proportion of development costs—almost 40 percent overall and 48 percent for those projects where this activity was performed. The next highest proportion was accounted for by bench-scale work, which averaged almost 24 percent of the total expenditure. (Thus, the gap between the first and second most costly activities is large.) After pilot-plant and bench-scale work, there is a substantial drop to about 12 percent, where we find market development, applications research, and preliminary engineering studies all clustered together. It appears from these statistics that toxicity studies, marketing research, and patent work play a very minor role in the development process in terms of total cost, at least in this sample of projects.

While these statistics are interesting, the reader must be careful to interpret them correctly. They are a measure of the central tendency of the proportions, *not* a measure of the proportional costs of developing the "average" product or

---

ucts. (See some of the case studies in Chapter 9.) Such costs are not included here because they are not required for the product's original development.

4. The projects were carried out by six firms, which we shall designate firms A to F. Firm A is a relatively small chemical firm, firm B is one of the largest chemical firms, firm C is a large oil firm, firm D is a major rubber company, firm E is one of the leaders in the chemical industry (and particularly in polymers), and firm F is a large oil firm.

5. If our sample does not contain a representative proportion of "small," "medium," and "large" projects, then not only would the average total cost of development be biased from the true mean for the population of polymer development projects, but also the relative costs of each activity may be biased. However, while the sample may be biased somewhat toward bigger projects, this bias does not seem very large or important.

TABLE 5.2

Average Proportion of Total Development Cost Attributable to Each Activity

	AVERAGE PROPORTION		
	ALL OBSERVATIONS	NONZERO OBSERVATIONS	PROPORTION OF NONZERO OBSERVATIONS
1. Preliminary engineering studies			
Mean	0.104	0.16	0.66
Standard deviation	0.161	0.18	
2. Toxicity studies			
Mean	0.000	0.004	0.04
Standard deviation	0.001	0.004	
3. Applications research			
Mean	0.12	0.18	0.66
Standard deviation	0.12	0.11	
4. Marketing research			
Mean	0.004	0.02	0.18
Standard deviation	0.009	0.01	
5. Patent and legal aspects			
Mean	0.01	0.04	0.34
Standard Deviation	0.03	0.04	
6. Bench scale			
Mean	0.24	0.30	0.79
Standard deviation	0.24	0.24	
7. Pilot plant			
Mean	0.39	0.48	0.82
Standard deviation	0.28	0.24	
8. Market development			
Mean	0.13	0.26	0.50
Standard deviation	0.22	0.26	

process. If, in each case, we were to weight the proportions for each development project by the cost of the project, we would obtain somewhat different results. These results are presented in Table 5.3, which shows that, for the "average" development project (which for this sample cost $376,245), almost 50 percent of total development cost was devoted to pilot plant, 19 percent to applications research, 14 percent to market development, 12 percent to bench-scale work, and 5 percent to preliminary engineering studies. Since preliminary engineering studies are not as well defined in the development process as are the other four activities, and since the weighted average proportion for this activity was only .05 (while the cumulative proportion for the other four activities was .93), it will be on the other four activities that attention will be focused in our econometric model described in section 7.

TABLE 5.3

Weighted Versus Unweighted Averages for the Proportion of Total Cost Attributable to the Five Most Costly Activities

ACTIVITY	AVERAGE	
	UNWEIGHTED	WEIGHTED
Preliminary engineering studies	0.10	0.05
Applications research	0.12	0.19
Bench scale	0.24	0.12
Pilot plant	0.39	0.47
Market development	0.13	0.14
Total [a]	*0.98*	*0.98*

[a] Because of rounding errors, figures may not sum to the total.

4. *Timing and Overlap of Activities*

Let's turn now to certain dynamic characteristics of development projects. Can a definite sequence be determined for the performance of the various activities? What is the nature of the overlap, if any, between the various activities? What factors seem to influence whether a particular activity is carried out?

The data examined thus far have shown the costs of each activity, including labor, overhead, capital, and material costs. In addition, manpower data—that is, the average number of men per month working on each activity—were also collected for each project in the sample. Since labor is not a homogeneous commodity, these data were further disaggregated to reflect the number of professionals and the number of technicians working on each activity.[6]

The advantages of using manpower data rather than cost data for this aspect of the investigation are two. First, there is the difficulty of allocating capital expenditures and expenditures for raw materials over time. If the construction of a pilot plant is required, it is not clear whether the costs of the pilot plant should be averaged over the duration of the operation of the pilot plant for the project, or whether some other accounting rule should be used. Any decision rule is very arbitrary. The second reason is that manpower data offer a convenient means of deflating our data for increasing costs of research and development over time. If manpower does comprise the major portion of development costs, as statistics would indicate, then the advantages of using these data may outweigh the disadvantages.

6. In terms of cost, it was assumed that one professional worker was equivalent to two technicians. This assumption was regarded as an appropriate (rough) approximation by most of the R and D project supervisors.

Ideally, what one would like for this investigation are concise data regarding the distribution over time of dollar costs and manpower requirements for each activity comprising the development project. However, accounting records generally do not permit the collection of accurate information of this sort. As a result, unless accounting records facilitated a different approach, the data were collected in such a way as to reflect the period over which a particular activity was performed, and the *average* number of technicians and professionals allocated to the project over this time period. While these data are less than ideal, they should be of use for the purposes at hand.

5. *Empirical Results Concerning Timing and Overlap*

Let's divide the total life of a project into ten equal periods, each being one tenth of the total time interval of the project. Figure 5.1 shows, for our sample of fifty-six development projects, the average amount of manpower expended in each activity during each such period as a percentage of the total manpower expended over the life of the project. For example, suppose that we consider a project which requires one hundred man-months of labor over the ten months which it takes to complete that project. Based on our fifty-six projects, Figure 5.1 tells us that, in the first month of that project, one would expect about 3.5 man-months of labor to be devoted to bench-scale work, 0.7 man-months to applications research, 0.8 man-months to pilot-plant work, 0.3 man-months to market development, and 4.2 man-months to the "other" four activities defined for this study.

From Figure 5.1, it is clear that, in the earlier stages of the development program, the bench-scale work is the dominant activity. However, the "other" activities, including preliminary engineering studies and the marketing-research studies, also comprise a fairly high proportion of the total development effort at this time. Once the 40 percent mark is reached in terms of total elapsed time, Figure 5.1 shows that the pilot-plant work begins to dominate the manpower needs of the development effort. This is followed by a rapid decline in the average amount of manpower devoted to the bench-scale work, and a steady rise in the amounts devoted to the pilot-plant and market-development activities. Over the last half of the project, approximately one half (on the average) of the manpower expended during each time interval is devoted to the pilot-plant effort, with one quarter to one third being devoted to the market-development effort and smaller amounts to the applications-research effort. To repeat, Figure 5.1 shows the average pattern for fifty-six projects; there is a great deal of variation of individual projects about this average pattern.

Clearly, although a definite sequence does seem to exist for three of the four major activities, the results indicate an overlapping of these activities. To measure this overlap, let us measure the time interval over which 20 percent to 80 percent of the labor input devoted to each activity was executed. Letting t_i be the

FIGURE 5.1

*Breakdown of Frequency Distribution of "Average" Manpower Effort, by Activity**

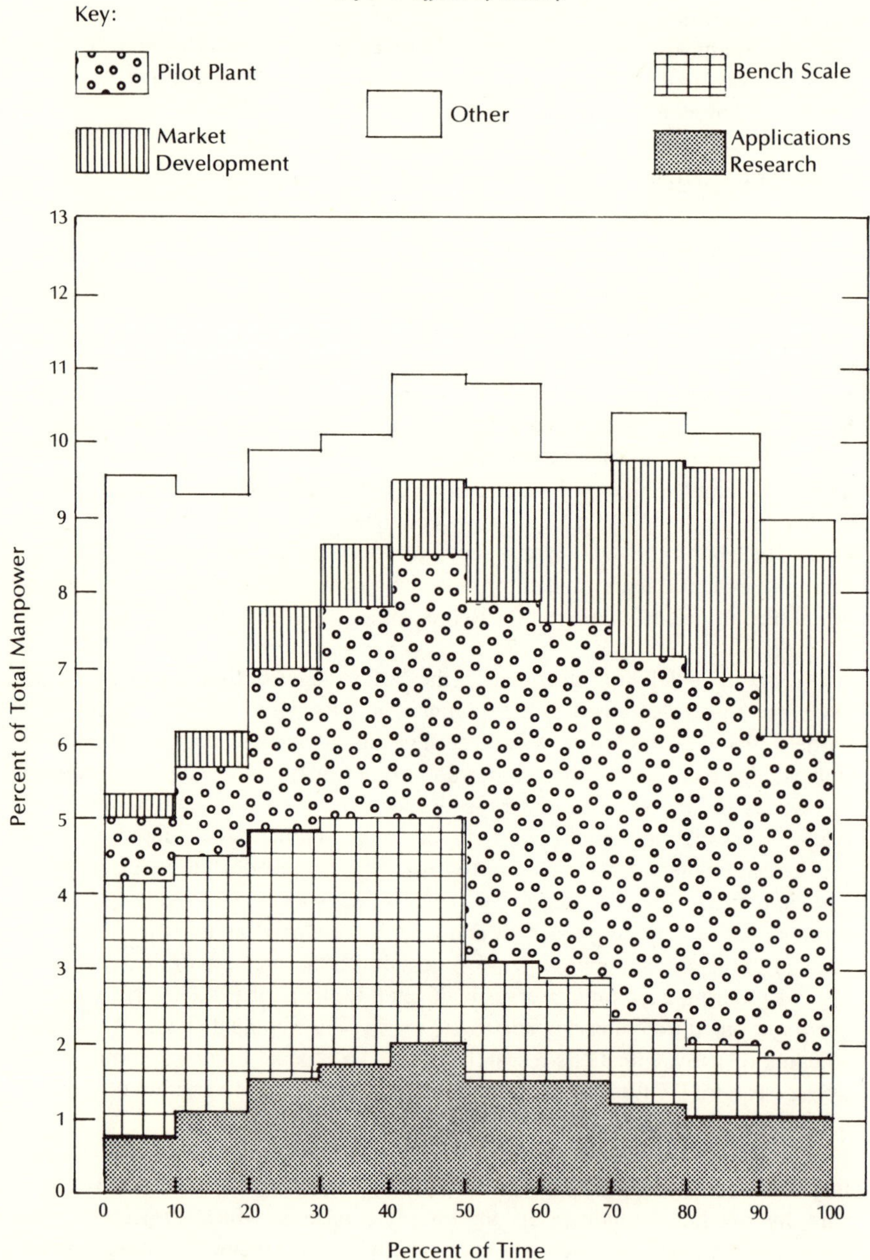

length of time over which 20 percent to 80 percent of activity i is performed, t_j be the length of time over which 20 percent to 80 percent of activity j is performed, and l_{ij} be the length of time which is shared in common by t_i and t_j, our measure of the degree of overlap is $O_{ij} = l_{ij}/(t_i + t_j - l_{ij})$. A value of zero for this statistic indicates that there is no overlap (by this definition) between the two activities, while a value of 1 indicates total overlap.

Based on this measure of overlap, let us examine the frequency with which the various activities overlap.[7] The results, shown in Table 5.4, indicate that in more than one half of the possible cases in which applications research can overlap with the other activities, it does. This confirms our impression that

TABLE **5.4**

*Frequency with Which Overlap Occurs between Various Activities**

	BENCH SCALE	PILOT PLANT	MARKET DEVELOPMENT
Applications research	0.56 (15)	0.61 (19)	0.53 (10)
Bench scale		0.28 (10)	0.30 (7)
Pilot plant			0.70 (14)

* The numbers in parentheses below the numbers in each cell of the table represent the number of projects in which the two activities overlapped. The main statistic represents this number as a proportion of the number of projects in which *both* of these activities were performed.

applications research, while a distinct activity, does not have a distinct niche in the timing of development projects, but is very likely to be performed concurrently with one or more of the other major activities.

Bench scale, on the other hand, presents an entirely different picture, for it appears that the major portion of the bench-scale work generally does not overlap with the other major activities (other than application research). In only ten out of thirty-six possible cases does the bench-scale overlap the pilot-plant work, and in only seven out of twenty-three possible cases does it overlap the market-development effort. Thus, the bench-scale phase of development does not appear to be characterized by much overlap with major activities other than applications research.

7. We must bear in mind that this analysis in no way assumes a consistent ordering of the activities. All we are trying to measure is the overlap of one activity with another without making any assumptions as to which activity commences first. There seems to be less consistency in the ordering of these activities than in the ordering of the stages of the innovation process (discussed in Chapter 4).

On the other hand, the pilot-plant and market-development activities seem to overlap relatively frequently, a fact which is easily explained by the interrelated nature of the two activities. For those projects for which a market-development effort is required, the close relationship between the pilot-plant operation (one purpose of which may be to produce material for the market-development effort) and the market-development effort is clear.[8]

6. *Type of Project and Activities Performed*

Obviously, there is a relationship between the activities performed in the development process and the type of development project which is carried out. Each of the projects in our sample was classified by the firm as a project designed to develop a new product, an improved product, a new process, or an improved process. Using these classifications, each of the projects was placed in the appropriate group and the activities performed in each project noted. In Table 5.5, we show the number of projects falling into each category, and the proportion of projects within each category requiring the execution of each of the individual activities. The results suggest that, in general, each type of development effort is associated with the execution of a somewhat different set of activities.

TABLE 5.5

Proportion of Projects Requiring Each Activity, by Type of Development Effort

TYPE OF DEVELOPMENT EFFORT	APPLICATIONS RESEARCH	BENCH SCALE	PILOT PLANT	MARKET DEVELOPMENT	TOTAL NUMBER OF PROJECTS IN CATEGORY
New product	0.80	0.84	0.88	0.84	25
Improved product	0.57	0.76	0.67	0.33	21
New process	1.00	1.00	1.00	0.00	4
Improved process	0.17	0.50	1.00	0.00	6

In particular, *new products* are generally characterized by execution of all four of the major activities, each of these activities being carried out in at least 80 percent of the projects. In contrast, *improved products* are generally characterized by concentration on the bench-scale and pilot-plant activities, each of which is executed in at least two thirds of the projects, with attention being paid to the

8. Our results indicate that there tends to be more overlap among high-priority projects that are carried out relatively quickly (relative to the minimum possible time) than among low-priority projects that are carried out relatively slowly (relative to the minimum possible time). This, of course, is what we would expect.

applications-research phase in about 50 percent of the projects and to the market-development effort more infrequently.

Turning to processes, each of the four projects concerned with the development of a *new process* required the execution of the applications-research, bench-scale, and pilot-plant activities. Not surprisingly, a market-development effort was not required for any of these projects. Finally, the six projects concerned with *improving an existing process* required the operation of the pilot plant, with only three requiring bench-scale work, one requiring applications-research work, and none requiring any market-development effort.

Thus, as one would expect, whether or not a particular activity is carried out depends on the nature and goals of the development project. In the next section,

TABLE **5.6**

Average and Standard Deviation of Cost and Duration of Development Projects, by Type of Development Effort

TYPE	COSTS (IN THOUSANDS OF DOLLARS)	TIME (IN MONTHS)
New product		
Mean	646	38
Standard deviation	1,481	26
Improved product		
Mean	86	18
Standard deviation	121	12
New process		
Mean	718	34
Standard deviation	632	14
Improved process		
Mean	42	16
Standard deviation	45	11

we go much further in discussing the determinants of the amount spent on each activity. An econometric model will be presented for this purpose. Before turning to that model, it is worthwhile looking at how the cost and duration of a project varies, depending on whether the project is aimed at a new product, a product improvement, a new process, or a process improvement. Table 5.6 shows that the average cost of a project aimed at a new process or a new product was about $700,000, and that the average duration was about three years. (Of course, for really major new products or processes, the cost and duration tend to be much greater.) On the other hand, the average cost of a project aimed at a product or process improvement was less than $100,000, and the average duration was about one and one half years. As indicated by the standard deviations in Table 5.6, there is an enormous amount of variation within each of these categories in the cost and duration of a project. Some of the factors causing this variation will be discussed in subsequent sections.

7. The Econometric Model: Overall Characteristics

Having carried out the analyses described in previous sections, we proceed to formulate an econometric model to explain and forecast the level of a project's costs of (1) applications research, (2) bench-scale work, (3) pilot-plant work, and (4) market development. An equation is estimated to explain each of these costs, this equation containing technical variables indicating the nature and extent of the "critical" problems investigated in this activity, as well as variables measuring other characteristics of the project and of the organization carrying it out. Then the technical variables are explained in turn by other variables characterizing the project and the organization carrying it out. Since the model is a recursive system, unbiased estimates of the parameters can be obtained by least-squares.

The equations explaining the costs of applications research and market development have much in common, as one might expect. In both cases, cost is higher if there is relatively great uncertainty concerning the specific needs or applications to which the product or process will be addressed, and if it is being developed for a market that is new to the firm. (Obviously, we are not implying that firms do not have in mind relatively well-defined applications for an innovation. In some cases, however, the uncertainties involved are much greater than in other cases.) Also, applications-research costs are influenced by whether or not a new plant must be built and by the number of fabricating techniques studied, which itself is dependent on a number of other variables. In the case of market development, costs are much higher, of course, if the project is a product development rather than a process development. Given the technical variables, these equations explain about one half of the variation in the cost of market development and about two thirds of the variation in the cost of applications research.

The equations explaining the costs of bench-scale work and pilot-plant work are more complicated. Bench-scale costs are determined by technical variables measuring the number of "critical" polymerization mediums tested and the sorts of other studies carried out, which are determined in turn by other characteristics of the project. The pilot-plant costs are influenced by technical variables measuring the number of "critical" unit operations studied in the pilot plant and the sorts of other studies carried out, which also are determined in turn by other characteristics of the project. An important variable is whether a new pilot plant has to be constructed, and some attempt was made to analyze the conditions under which this is the case. Given the technical variables, these equations explain about two thirds of the variation in bench-scale cost and about four fifths of the variation in pilot-plant costs.

8. The Detailed Results for Applications Research

Given the limitations of space and the fact that a detailed description of the entire model is available elsewhere,[9] we shall not try to describe all of the equations in the model taken up in the previous section. But to indicate more specifically the types of variables included and the sorts of results obtained, we shall present a detailed discussion of the results concerning one of the equations—the one pertaining to the cost of applications research. This should indicate the general spirit and structure of the model in enough detail for present purposes.

In essence, the function of the applications-research activity is to test a polymer for various applications or uses. In order to accomplish this, the various properties of the polymer must first be analyzed and its characteristics charted. Then the applicability of the polymer in various uses must be examined, with careful attention being paid to the products (if any) with which this polymer must compete. As indicated earlier, the tests of the polymer will try to ascertain its performance characteristics as well as its ease of fabrication. The ultimate cost of this phase will depend on the number of tests which must be performed and the cost of each test.

In constructing our model of the cost of applications research, we would expect the following variables to be important determinants of this cost:

(1) *Development to meet a specific need* (*NEED*). The cost of the applications-research activity depends heavily on the extent of the uncertainty at the outset of the project concerning the specific needs or applications to which the innovation will be addressed. To see what we mean, consider two polar cases, each of which is presented for illustrative purposes, not because it is regarded as typical. In the first case, there is a great deal of uncertainty at the outset of the development project regarding the specific end use for the product. Nonetheless, because of a number of attractive properties of the polymer, the project is started. At the other extreme is the situation in which the development project simply involves a modification or improvement in a product at the request of a customer. In this case, there is practically no uncertainty concerning the specific end use of the innovation. It is quite evident that the former case is likely to dictate many more tests and more expenditure on applications research than the latter.

In order to measure this effect, the following question was posed to the R and D managers: "Although this is an oversimplification, it is possible to distinguish between products which are developed when there is considerable uncertainty concerning the particular set of uses at which they will be directed, and products which are developed to meet a rather specific need. Into which of these categories would you classify this product?" A dummy variable was then constructed which assumed a value of 1 if the product was developed for a specific

9. See E. Villani, "Determinants of Development Costs in the Chemical Industry: An Economic Investigation" (Ph.D. diss., University of Pennsylvania, 1973).

need and a value of zero if it was not. This variable would be expected to have a negative coefficient in the equation to explain the costs of application research.

(2) *New Market* (*M*). Another important factor influencing the cost of applications research is the extent of the firm's familiarity with the markets for which the product is being developed. (A market is defined in the paragraph after next.) We would expect that a product which is being developed for some markets which are unfamiliar to, or at least relatively unexplored by, the firm would entail more testing in applications research. One reason for this is that in such a situation the firm is just not familiar with the performance characteristics required of its product in such markets, or, in any case, might not be equipped to test these characteristics. In order to test for this effect, the variable *M*—which is 1 if any of the relevant markets is new, zero otherwise—will be included in the equation to explain the costs of applications research, and will be expected to have a positive coefficient.

(3) *Number of markets and customers* (*MARK, CUST*). Although a discussion of the novelty of the markets for which the product is intended has been included, one set of factors which has not been included is the number of markets for which the product is being developed and the number of potential customers within each market. One might reasonably expect that the amount of testing which would be required in applications research would be a function of both of these variables,[10] for it is reasonable to assume that each market might require specific sets of tests not required for other markets, and that the same might be true for each customer.

In order to capture the influence of these two factors, we determined in each case the number of markets for which the product was developed and the number of customers personally contacted within each market. A market was defined for purposes of this question as a five-digit industry (as defined by the Standard Industrial Classification). Two variables were constructed, the first of which, *MARK*, represented the number of markets, and the second of which, *CUST*, represented the total number of potential customers for which the product was being developed.[11]

10. The combination of these two variables would be analogous to the variable which Schnee refers to as the ''spectrum of activity'' in his study of a large ethical pharmaceutical company. Schnee used this variable to refer to the range of a drug's biological activity. The rationale was that the broader the spectrum of activity of a drug, the greater the number of therapeutic markets to which it was applicable. In each case where a product is introduced into two or more distinct therapeutic markets, the efficacy and safety of the drug must be established in the separate markets. See E. Mansfield, J. Rapoport, J. Schnee, S. Wagner, and M. Hamburger, *Research and Innovation in the Modern Corporation* (New York: Norton, 1971).

11. By potential customer, we mean a customer which had been contacted personally and which the firm believed to be very likely to buy the product when commercialized.

(4) *Number of fabricating techniques studied* (*FAB*). In previous sections, we have stressed that applications research entails the testing of the various properties of the product under specified conditions. However, there is another side of the applications-research phase which must not be neglected, namely, the fabrication of the final product. In many instances, a great deal of time and effort must be devoted to such fabricating procedures. Since it is likely to be a highly capital-intensive procedure, possibly involving a large outlay for capital equipment, it may lead to considerably added expenditure on applications research.

Hence, for each development project, the R and D manager was asked to specify which and how many fabricating techniques were tested in applications research. From the answers to this question, the variable, *FAB*, was constructed. This variable, which simply measures the number of different fabricating procedures tested, is the only aggregated technical-activity variable used for the applications-research activity. While it contains all of the limitations of such an aggregate variable, it should provide insight into the scope of the technical activities performed. *FAB* is, of course, an endogenous variable in our entire model, its determinants being taken up in note 15.

(5) *Existence of final production unit* (*PLANT*). An important consideration in any development program is the final production unit for the product once (and if) the product is commercialized.[12] A development project often is more attractive if the end product can be produced in existing plant facilities. This is, in fact, the case with most of our development projects.[13] However, if the construction of a plant is required, this may involve added expenditure for the applications-research activity, since considerably more attention must be paid to the fabricating techniques and to the testing of the product. For this reason, the variable, *PLANT*, which assumes a value of 1 for projects requiring the construction of a new plant and zero otherwise, was included as a determinant of the cost of applications research.

(6) *Calendar year* (*YEAR*). No consideration has been given as yet to the possibility that the costs of applications research have changed systematically over time. One major reason why the calendar year might be expected to have a positive effect on development costs is that there has been a general rise in the

12. A possible point of confusion should, perhaps, be clarified. The sample includes a wide spectrum of projects, ranging from the simple improvement of an existing plastic product to the development of a totally new polymer. Thus, in speaking of the final production unit for the product, we may either be speaking of a plant to produce a polymer in its raw form (e.g., pellets or powder) or a plant in which the output is a final consumer product (e.g., a polyethylene bag or nylon hose). In the latter case, the study of the product's properties and of the various fabricating techniques for the product is likely to play a more important role.

13. Forty-six of the fifty-six projects in the sample did not require the construction of a new plant.

price deflator for research and development over time.[14] Hence, *ceteris paribus,* one would expect the cost of applications research to rise over time due to the increases in equipment and material costs, and the rising salaries of scientists, technicians, and supporting personnel. Since the fifty-six projects included in this sample were initiated over a long span of years (1954 to 1970), this factor should be taken into account. The variable, *YEAR,* was constructed by taking the calendar year in which the project was initiated and subtracting from it 1954, the year in which the first project was initiated.

9. *Economic Results: Applications Research*

Various regressions were estimated, using the variables described above as the independent variables. The equation to be presented below is the best of those which were estimated. Neither *MARK* nor *CUST* has been included in this equation since neither was found to be statistically significant. Even though there is a certain degree of collinearity between *MARK* and *M,* it is quite obvious that when *MARK* is substituted for M they are measuring two different phenomena. The apparent conclusion is that, while a new market significantly affects the expenditure devoted to applications research, the number of markets (and/or the number of customers) does not significantly influence these costs for the projects in this sample. The variable, *YEAR,* was not included in this equation since it too was found to be statistically nonsignificant. Either the increased efficiency of the firms in the sample has balanced the increasing cost of doing R and D over time, or an implicit deflator is contained in the other variables in the equation (e.g., the firm dummy variables).

The equation chosen for our model was the following:

$$\begin{aligned}(5.1)\quad AR = {} & \underset{(2.37)}{41.71\,FAB} + \underset{(2.75)}{153.60\,PLANT} - \underset{(-2.44)}{125.56\,NEED} + \underset{(1.58)}{64.40\,M} \\ & + \underset{(0.70)}{42.80\,F_A} + \underset{(0.70)}{35.13\,F_B} + \underset{(3.82)}{394.09\,F_C} - \underset{(1.32)}{212.97\,F_D} \\ & + \underset{(1.27)}{115.95\,F_E} + \underset{(4.06)}{382.15\,F_F},\end{aligned}$$

where AR = total dollar expenditures for applications research (in thousands of dollars); FAB = number of fabricating techniques studied in applications re-

14. See E. A. Johnson and H. S. Milton, "A Proposed Cost of Research Index," *IRE Transactions on Engineering Management* (December 1961). The authors find that the cost of doing research and development work was 91 percent higher in 1960 than in 1950. Also see J. Jewkes *et. al., The Sources of Invention,* rev. ed. (New York: Norton, 1970). The authors note the dramatic rise in the cost of development over the past century.

search; *PLANT* = 1 if construction of plant was necessary to produce final product, 0 otherwise; *NEED* = 1 if the project was a product development and if the product was developed to meet a very specific need, 0 otherwise; $M = 1$ if any of the product's markets was new to the firm, 0 otherwise; $F_i = 1$ if project originated in firm i, 0 otherwise. (The six firms in our sample are labeled A, B, C, D, E, and F.)

The results explain much of the observed variation in the cost of applications research, the coefficient of determination ($\bar{R}^2$), adjusted for degrees of freedom, being 0.67. Since the equation has been specified as a linear relationship, it implies that the incremental cost attributable to each variable in the equation is a constant fixed sum across projects. An alternative assumption would have been that each variable has a multiplicative effect on cost. To test for this possibility, a logarithmic form was estimated for this equation (as well as for the equations for the other three major activities) and was found to provide an inferior fit both in terms of the statistical significance of each of the coefficients and the total variance explained by the equation.

Perhaps even more critical than the assumption of a linear relationship is the fact that so many qualitative (i.e., zero–1 or dummy) variables are included in our equation. Although this may not present any problems from a statistical standpoint (as long as these dummy variables are truly exogenous), we do run the risk that the equation may be misspecified, since by including these variables in this manner, we are making the implicit assumption that each one of the respective qualitative factors affects only the intercept of our equation—i.e., we are assuming that the effect of the respective variable is to increase (or decrease) the cost of the activity by a constant amount in each project, with this amount independent of the other factors influencing the cost of the development project. This is obviously a strong assumption, and not likely to be strictly true for every project. The alternatives would have been: (1) to include various cross products of these qualitative variables and to test for their statistical significance; or (2) to group the data by the appropriate classifications, estimate an equation for each, and then test whether the coefficients of the equation were significantly different across classifications. Given the relatively small size of the sample, however, it was not feasible to carry out much experimentation of this sort along either of these lines.

Before discussing the numerical values of the coefficients in the equation, we must stress that each of these coefficients represents the "average" effect of a particular variable on the cost of applications research. While the coefficients are generally statistically significant, it must be recognized that there is likely to be a great deal of variation around the "average" in the case of any particular development project. For example, the coefficient of *FAB* in the above equation is positive and statistically significant. (The t ratios are given in parentheses below the regression coefficients in equation (5.1), and subsequent equations.) The interpretation of this coefficient is that each fabricating technique tested for the projects in our sample required, on the average, an expenditure of about \$40,000. However, it should be evident that, for any particular development project, this figure is likely to depend on a number of other factors associated with the development program, and that the actual cost of studying each fabricating technique may be more or less than the above figure, which is only an average. This caveat should be borne in mind throughout this section.

All of the regression coefficients have the hypothesized signs. The coefficient of *PLANT* (which is statistically significant) tells us that, *ceteris paribus,* expenditures for applications research increased an average of about \$150,000 (for the projects in our sample) when a project required the construction of a new plant. The coefficient of *NEED* (which is also statistically significant) indicates that the applications-research costs of projects where there is considerable uncertainty concerning end use are (for the projects in our sample) about \$125,000 higher, on the average, than those of projects which are established to satisfy a very specific need. Finally, the coefficient of *M* is positive, but not statistically significant at the 5 percent level. A likely reason for this is its collinearity with *FAB*. While this detracts from its statistical significance, it was kept in the equation because, on a priori grounds, it seemed very unlikely that it did not affect AR.

An examination of the coefficients for the six firm variables reveals that, for four of these firms, these coefficients do not differ significantly from zero. However, for firms C and F, the coefficients are significantly different from zero and quite large. Four of the six projects originating in these two firms required the construction of new plants. However, even the inclusion of the variable, *PLANT,* while considerably reducing the magnitude of these two coefficients, does not eliminate their statistical significance. Although very difficult to substantiate empirically with the data now at our disposal, we believe (on the basis of discussions with the R and D people participating in these projects) that the statistically significant positive coefficients for firms C and F may be attributed to (1) the limited previous experience of these firms in this area, (2) the fact that part of the costs they attributed to this phase might have been more correctly designated as research, and (3) the relative inefficiency with which these firms performed these projects.[15]

10. Forecasts of Development Costs

Our econometric model contains equations for four (of the eight) activities which together account, on the average, for approximately 93 percent of the total

15. As noted above, *FAB* is an endogenous variable. To complete the model for applications-research cost, an equation was formulated to explain *FAB*. Among the variables used to explain *FAB* were *M, YEAR,* the number of customers for which it was necessary to demonstrate the innovations in the customer's plant, and whether or not the innovation was a new product. For this equation, $\bar{R}^2$ was 0.59, and all of these variables were statistically significant.

cost of development. Therefore, ignoring the variation in this percentage, since it is small, we can estimate the total cost of development by using the equations in the model to estimate *AR, BS, PP,* and *MD,* where *AR* is the cost of applications research, *BS* is the cost of bench scale, *PP* is the cost of a pilot plant, and *MD* is the cost of market development. Then we can insert these estimates into the following formula:

$$\text{Total cost} = (AR + BS + PP + MD) \div 0.93 \tag{5.2}$$

Using this model, how well can we "forecast" development costs? To find out, we computed all of the equations in the model, omitting the five most recent projects in our sample. In other words, we act as if we had only the data available before these five projects were carried out. Then we use the resulting equations to forecast AR, BS, PP, and MD for each of these five projects.[16] The resulting forecasts were then inserted into equation (5.2) to obtain forecasts of total costs, and the accuracy of the results was compared with the accuracy of the firms' actual forecasts.[17]

The comparison reveals that the model-generated forecasts do significantly better than the actual forecasts made by the firms themselves. The root-mean-square-deviation-from-1 of the ratio of actual to estimated costs for the model-generated forecasts was 0.60, while it was 1.00 for the firms' forecasts. Certainly, these results are encouraging. Although the model is rough, it seems to forecast reasonably well, at least when judged by this criterion. Also, if one wants to make forecasts based on the NSF definition of development, it is a simple matter to do so, since $(AR + \mathrm{BS} + PP)$ is a good approximation to this measure. Thus, one can use the equations for *AR, BS,* and *PP* to make such forecasts. The forecasting accuracy of the model seems about the same, regardless of which definition of development is used.

16. In any case where the forecasted value of *AR, BS, PP,* or *MD* was negative, this value was set equal to zero. This procedure, although rough, seems to be reasonable since none of these costs can be negative.

17. On the average, actual development cost was about 31 percent greater than the firms' forecast. The standard deviation of the ratio of actual to forecasted cost (by the firms) was 0.54. The cost overruns tend to be smaller if the project is a spin-off from previous work conducted by the firm, and if the primary problems are regarded as technical rather than in marketing. They tend to be larger if the project is a product development (rather than a process development), if parallel efforts were used, and if the product or process was developed for a single customer. On the average, the actual duration of a development project was about four times the duration estimated by the firms. And the standard deviation of the ratio of the actual duration to the forecasted duration was 7.7. Relative to our previous results for other industries, the cost estimates seem relatively accurate (as these estimates go), but the time estimates seem relatively poor. See. E. Mansfield *et al., Research and Innovation, op. cit.*

Finally, we must emphasize that this model, as it stands, is not put forth as an operating tool. It is viewed as an experimental device to help indicate the applicability of econometric techniques in this area. Without question, it is crude, as is generally the case with initial attempts of this sort. All that we are willing to conclude is that the results seem encouraging.

11. Summary and Conclusions

This chapter attempts to add to the small body of economic literature dealing with the development process within the firm. Employing the classification of activities described in the previous chapter, we obtained very detailed data concerning fifty-six polymer development projects. The results show that, on the average, 93 percent of the development costs were accounted for by the pilot-plant, bench-scale, applications-research, and market-development activities. The costs of the pilot-plant activity accounted, on the average, for almost one half of the total development costs, while the applications-research, bench-scale, and market-development activities each accounted for about 15 percent.

During the early stages of a development project, the bench-scale activity tends to be the dominant one, while during the last half of a project, the pilot-plant and market-development activities tend to be dominant. There is often considerable overlap between the pilot-plant and market-development activities. Also, applications research frequently overlaps with many of the other activities. Whether or not a particular activity is carried out depends on whether a project is aimed at an entirely new product, a product improvement, a new process, or a process improvement. Projects aimed at entirely new products generally contain all four of the major activities. On the other hand, projects aimed at product improvements seldom have a market-development phase (and frequently there is no applications research). Projects aimed at new processes seem to contain all activities other than market development, whereas those aimed at process improvements generally have no applications research (and frequently there is no bench-scale work either).

We formulated an econometric model to explain and forecast the level of a project's costs of (1) applications research, (2) bench-scale work, (3) pilot-plant work, and (4) market development. An equation was estimated to explain each of these costs, this equation containing technical variables indicating the nature and extent of the "critical" problems investigated in this phase, as well as variables measuring other characteristics of the project and of the organization carrying it out. For example, the cost of applications research depends on whether or not the product was developed to meet a very specific need, on whether any of the product's primary markets was new to the firm, on the number of fabricating techniques studied, and on a number of other variables.

Then we used this model to forecast the total development costs of five projects. The results were quite encouraging, since our forecasts were more accurate than those of the firms. The root-mean-square-deviation-from-1 of the ratio of actual to estimated costs was 0.60 for our forecasts and 1.00 for the firms' forecasts. Needless to say, this does not mean that our model is anything more than a very rough first approximation. Clearly, many of the variables are crude, and simplifications of many kinds had to be made. This model should be viewed solely as an experimental effort, not as a practical tool. But this model is one of the first of its kind, there having been very few econometric studies of the determinants of development costs; and we hope that it may be a useful beginning.

6 DETERMINANTS OF THE SPEED OF APPLICATION OF NEW TECHNOLOGY

1. Introduction

Up to this point, we have been concerned primarily with the process by which firms produce new technology and embody it in innovations—that is, new and improved products and processes. In this chapter, we investigate how rapidly such innovations spread. To discuss this topic, which is of the utmost importance for both economic analysis and policy, we must shift our attention from the individual firm to entire industries and the economy as a whole. Besides trying to summarize briefly what is known in this area, we present the results of some new econometric studies of the rates of diffusion of innovations in the chemical and machine tool industries. These results break new ground, since they indicate how useful various models have been in forecasting and how well existing models can be extended to cover new sets of circumstances. Also, we present for the first time an analysis of the diffusion of product (as distinct from process) innovations.

2. The Rate of Diffusion of Industrial Innovations

Once a new technology becomes available, it takes time for it to be commercialized, and then it takes more time for the diffusion process (described in Chapter 1) to run its course. In the United States, how rapidly does the diffusion process go on? According to the available data, it takes about 5 to 10 years, on the average, before one half of the major firms in an industry begin using an important innovation. And in many cases it takes longer. The rate of imitation varies widely: Although it sometimes takes decades for firms to install a new technique, in other cases they follow the innovator very quickly. For example, it took about fifteen years for half of the major pig-iron producers to use the by-product coke oven, but only about three years for half of the major coal producers to use the continuous-mining machine.

Even when a firm begins using a new technique, this does not mean that the diffusion process is over for this firm. It generally takes a number of years before a firm completes the substitution of the new technique for the old. For example, once it began to dieselize, it took about nine years, on the average, for an American railroad to increase its stock of diesel locomotives from 10 percent to 90 percent of the total. And there was wide variation among firms in the intrafirm rate of diffusion. For example, some firms took only a few years to go from 10 to 90 percent, whereas others took fourteen years or more.[1]

3. A Simple Model of the Imitation Process

What determines how rapidly the use of a particular new process spreads from one firm to another? Given that one firm has begun using a new technique, what determines how rapidly other firms begin using it? Some years ago, one of the authors suggested that the following simple model might be of use:[2] Letting $\lambda_{ij}(t)$ be the proportion of firms in the ith industry not using the jth innovation at time t that introduce it by time $t + 1$, he proposed that

$$\lambda_{ij}(t) = f_i(P_{ij}(t), \pi_{ij}, S_{ij}, \ldots), \tag{6.1}$$

where $P_{ij}(t)$ is the proportion of firms in this industry that have introduced it at time t, π_{ij} is the profitability of installing this innovation relative to that of alternative investments, and S_{ij} is the investment required to install this innovation as a percent of the average total assets of the firms. In other words, the model assumes that the probability that a nonuser will use the innovation between time t and $t + 1$ is dependent on the proportion of firms already using the innovation,

1. E. Mansfield, *Industrial Research and Technological Innovation* (New York: Norton, 1968). For a study of the growth over time of the percent of output produced by a sample of major innovations, see E. Mansfield, "The Diffusion of Eight Major Industrial Innovations in the United States," in N. Terleckyj, (ed.) *The State of Science and Research: Some New Indicators* (Washington, D.C.: National Planning Association, 1976). Also, see J. Tilton, *International Diffusion of Technology: The Case of Semiconductors* (Washington, D.C.: Brookings 1971), and J. Utterback, "Innovation in Industry and Diffusion of Technology", *Science* (February 15, 1974).

2. E. Mansfield, "Technical Change and the Rate of Imitation," *Econometrica* (October 1961).

the profitability of using the innovation, and the investment required to install the innovation.

Certainly, it seems reasonable that this probability will increase with increases in $P_{ij}(t)$, the proportion of firms already using the innovation. The larger the proportion of the firms already using the innovation, the less risky it is for a nonuser to begin using it, and the more competitive pressure there may be on a nonuser to begin using it. Also, there tends to be a "bandwagon" effect. Turning to the effects of π_{ij}, it seems eminently reasonable that more profitable innovations will tend to be accepted more rapidly. Also, one would expect that increases in S_{ij}, the investment required to introduce the innovation (as a percent of the average total assets of the firms), would result in a slower rate of diffusion.

Assuming that $\lambda_{ij}(t)$ can be approximated adequately by a Taylor's expansion that drops third and higher-order terms, and assuming that the coefficient of $P_{ij}^2(t)$ in this expansion is zero, it can be shown that the growth over time in the number of firms having introduced the innovation should conform to a logistic function. Specifically,

$$P_{ij}(t) = [1 + e^{-(L_{ij} + \phi_{ij} t)}]^{-1}. \quad (6.2)$$

It can also be shown that the rate of imitation depends only on ϕ_{ij}, and on the basis of our assumptions,

$$\phi_{ij} = b_i + a_1 \pi_{ij} + a_2 S_{ij} + z_{ij}, \quad (6.3)$$

where the *a*s and *b*s are parameters and z_{ij} is a random error term.

This model has been tested against data for over a dozen innovations in five industries, the results being quite favorable. In general, the growth in the number of users of an innovation can be approximated by a logistic curve. And there is a definite evidence that more profitable innovations and ones requiring smaller investments had higher rates of imitation, the relationship being similar to that predicted in equation (6.3).[3] Moreover, Hsia has found that this model provides a good fit to data regarding twenty-six innovations in the plastics, textiles, and electronics industries in Hong Kong. And Blackman has found this model to be useful in his studies of the U.S. aircraft-engine industry.[4]

3. *Ibid.* Also, see Z. Griliches, "Hybrid Corn: An Exploration in the Economics of Technological Change," *Econometrica* (October 1957), for an excellent study of the diffusion of an agricultural innovation.

4. R. Hsia, "Technological Change in the Industrial Growth of Hong Kong," in B. R. Williams, ed., *Science and Technology in Economic Growth* (New York: Macmillan, 1973); and A. W. Blackman, "The Rate of Innovation in the Commercial Aircraft Jet Engine Market," *Technological Forecasting and Social Change* (1971). For additional relevant and interesting data, see L. Nabseth and G. Ray, *The Diffusion of New Industrial Processes* (London: Cambridge University Press, 1974). In general, their results seem to support the hypotheses underlying this model.

4. Application of the Model: Forecasting the Rate of Diffusion of Numerically Controlled Machine Tools

Although this model is oversimplified in many respects, it has proved useful in forecasting the rate at which the use of particular innovations would spread. For example, it was employed in a study carried out for the Small Business Administration at the beginning of 1968 to forecast the percent of firms in the tool and die industry that would be using numerically controlled machine tools in 1970. When the study was carried out, about 20 percent of the firms in the National Tool, Die, and Precision Machining Association were using numerical control. Two sets of data were obtained: interview data and mail-survey data. Based on the interview data, the model forecasted that about 33 percent of the firms would be using numerical control at the beginning of 1970. Based on the mail-survey data, the model forecasted that about 37 percent of the firms would be using numerical control at the beginning of 1970.[5]

To see how these forecasts compared with those obtained on the basis of other methods, we constructed two alternative types of forecasts. First, we asked the firms—both in interviews with a carefully selected sample of firms and in the mail survey of the industry—whether they planned to begin using numerical control in 1968 or 1969. Since there was considerable lead time required in obtaining numerical control, it seemed reasonable to suppose that their replies would have some forecasting value. The results of the interviews indicated that about 16 percent of nonusers planned to use numerical control by 1970; the results of the mail survey indicated that this was the case for 28 percent of the nonusers. Thus, the forecast was 33 percent, based on the interview data, or 43 percent, based on the mail survey. Second, we obtained forecasts from the machine tool builders, the firms that presumably are closest to and best informed about the market for numerically controlled machine tools. About twenty-five of the 150 members of the National Machine Tool Builders Association provided forecasts. The result showed a considerable amount of variation, but the median forecast was about 30 percent.

How accurate were these forecasts? Which forecasting approach was most accurate? Table 6.1 shows that the model's forecast based on the data from the mail survey was almost precisely correct and that the model's forecast based on the interview data was off by only 4 percentage points. Regardless of whether we look at results based on the interview data or the mail survey, the model forecasts

5. The mail survey resulted in data for about three hundred firms. The interview data were based on a carefully designed two-stage sample of all firms. In both cases, the model was used to forecast the 1970 figure on the basis of the time series up to 1967. However, it was not appropriate to assume, as does equation (6.2), that the upper asymptote is one. Instead, two projections were made, one based on an upper asymptote of 0.5, and one based on an upper asymptote of 0.8. The midpoint of this range of projections is the forecast we consider here. See E. Mansfield, *Numerical Control: Diffusion and Impact in the Tool and Die Industry,* report to the Small Business Administration, 1968.

better than the other two techniques. Moreover, it forecasts better than simple extrapolation by the usual "naïve" models.[6] Certainly, this is encouraging. Based on these and other results, it appears that this simple model may be of use in forecasting the rate of imitation. Of course, this does not mean that it is anything more than a crude device—or that it can be applied in situations where its basic assumptions do not hold. But it does mean that, used with caution, it may perform at least as well as other commonly used forecasting devices.[7]

TABLE **6.1**

Alternative Forecasts (Made in 1968) of the Percentage of Firms in the U.S. Tool and Die Industry Using Numerical Control by 1970, and the Actual Percentage

TYPE OF FORECAST	BASED ON INTERVIEW DATA	BASED ON MAIL SURVEY
	(percent)	
Model	33	37
Plans of tool and die firms	33	43
Median forecast by machine tool builders	30	30
Actual percentage[a]	37	37

[a] See note 7.

5. *Extensions of the Model*

Although the basic model outlined in section three is sufficient for some purposes, it omits many important variables. This section indicates several directions in which this model can be—and is being—extended. First, science-based industries that spend a great deal on research and development may tend to have higher rates of imitation, holding other factors constant, than industries that are based on technologies more remote from science and that do little R and D. This hypothesis (which we shall test in Chapter 7) seems reasonable, since research-

6. Specifically, the model forecasted better than naïve models that assumed that the increase in the percentage of firms using numerical control would be the same amount, in absolute or relative terms, during the next two years (1968–1969) as it had been during the previous two years (1966–1967).

7. To estimate the actual figure for 1970, we assumed that the ratio of the national percentage of firms using numerical control to the percentage of firms in the Philadelphia region using numerical control was the same in 1970 as it was during 1965–1967. This assumption seems reasonable, and this is the best estimate one can make. The NTDPMA directory is unreliable in this respect in 1970, according to A. Romeo's results.

intensive industries are better able to evaluate new technologies and are likely to put up less resistance to change than industries that do relatively little R and D.

Second, an industry's market structure may have an effect on the rate of imitation. Some economists believe that innovations spread more rapidly in more highly concentrated industries. Our own previous studies do not bear this out, but the results are based on only a limited sample. In Chapter 7, we shall extend the model to include measures of industrial concentration, and present new econometric findings that bear on the effects of an industry's market structure on the rate of diffusion.

Third, the dispersion of the profitability of the innovation among firms is likely to have significant effects on the rate of imitation. Holding the average profitability of the innovation constant, the rate of imitation is likely to decrease as the variance among firms in the profitability of the innovation increases. For example, suppose that, on the average, firms can get a 30-percent return from a new process. If there is wide variation of firms about this average, the firms where the rate of return is much lower than 30 percent will be much slower to adopt the new process than the firms where the rate of return is much greater than 30 percent. On the other hand, if there is little variation of firms about the average, there is likely to be less variation in how rapidly firms begin using the new technique, with the result that the rate of imitation will be higher. Some attempt is currently being made to test this hypothesis.

Fourth, another factor that can have an important influence on the rate of imitation is the length of time that the innovation has been in use in other industries. Many innovations come to be used in a variety of industries; for example, numerically controlled machine tools are now used in the aircraft, automobile, electrical equipment, farm machinery, instrument, printing press, tool and die, and many other industries. Since the information and experience that is gathered in one industry can help to resolve uncertainties regarding the innovation in other industries in which it is introduced subsequently, one would expect the rate of imitation in an industry to be directly related to the length of time that the innovation has been in use in other industries (before this industry begins using it). Some attempt is currently being made to test this hypothesis too.

6. Application of the Model to Output Data in the Chemical Industry

Another important way we have attempted to extend our basic model is to look at other measures of the rate of diffusion of an innovation besides the rate of imitation. For many purposes, it is more important to know how rapidly the percent of output produced by the new technique increases, or to know how rapidly the percent of new equipment that embodies the new technique increases, than to know how rapidly the percent of firms using the new technique increases. In the next chapter, we shall look at the rate of diffusion, as measured by the increase over time in the percent of new equipment embodying the new technique. In this section, we shall see how well the basic model can be extended to cover the increase over time in the percent of output produced by a new technique. Specifically, we summarize in this section results obtained by Simon, in his work as part of our project, concerning the diffusion of new processes in the chemical industry.[8]

Let us redefine $\lambda_j(t)$ as the proportion of industry output not produced by the jth process innovation at time t that is produced by it at time $t+1$. (Since we are concerned here only with the chemical industry, the i subscript can be omitted.) Also, let's redefine $P_j(t)$ as the proportion of industry output produced by the jth process innovation at time t. Then if we make the assumptions set forth in section 3, we obtain the same expressions as in equations (6.3) and (6.4), but the definition of $P_j(t)$ is different, the diffusion rate now being measured by the increase over time in the percent of output produced by the new technique. As before, the diffusion rate depends only on ϕ_j, which can be estimated by regressing $\ln [P_j(t) \div (1 - P_j(t))]$ on t.

Simon obtained detailed data concerning the growth over time in the percent of output produced by each of twenty-three process innovations in the chemical industry. Based on the regression of $\ln [P_j(t) \div (1 - P_j(t))]$ on t, he obtained the estimate of ϕ_j and L_j for each innovation shown in Table 6.2. In accord with our basic model, the expression in equation (6.2) seems to fit quite well for most of these innovations. Table 6.2 shows that $\bar{R}^2$ (corrected for degrees of freedom) is high in all but a few cases.[9]

In the chemical industry, there is good reason to believe that innovations tend to spread more rapidly in recent years than in the more distant past, because of the growth of more effective mechanisms to transmit and evaluate technical information. For example, the engineering literature is more abundant and detailed, and evaluation techniques are more sophisticated. Also, it seems reasonable to believe that, in cases where other firms could and did license the innovator's process, the innovation would spread more rapidly than in cases where the imitators had to develop their own versions of the process.

If these two variables are also included as determinants of λ_j in equation (6.1), it turns out that they are also among the independent variables in equation (6.3). Using least-squares estimates, Simon finds that

8. P. Simon, "Models of Process Diffusion and Entry in the U.S. Chemical Industry" (Ph.D. diss., University of Pennsylvania, 1975). In the available space, it is impossible to provide more than a very incomplete, and preliminary account of Simon's findings. He will publish a fuller description of his work at a later date.

9. Simon includes only the period when the process was dominant, and assumes that the upper asymptote of the growth curve was 100 percent. Two of the innovations did not spread, so their values of ϕ_j were set equal to zero. Obviously, it would have been inappropriate to estimate ϕ_j in the usual way in these two cases.

TABLE 6.2

Estimates of Parameters, Twenty-three Chemical Process Innovations

INNOVATION	$\hat{L}_j$	$\hat{\phi}_j$	$\bar{R}^2$
Acetaldehyde—Wacker process	−2.43 (−11.99)	0.325 (12.12)	0.98
Acetic acid—LPG oxidation (see note 9)	n.a.	0	n.a.
Acetylene from hydrocarbons	−2.459 (−13.39)	0.136 (9.10)	0.83
Acrylonitrile—Sohio process	−3.252 (−15.53)	0.559 (18.09)	0.97
Ammonia—single-train plants, 600 ton/day or larger	−3.386 (−4.88)	0.431 (4.09)	0.74
Aniline by catalytic hydrogenation of nitrobenzene	−1.586 (−2.78)	0.280 (4.10)	0.85
Butadiene from butane	−1.813 (−4.54)	0.066 (3.59)	0.52
Carbon disulfide from methane	−4.494 (−2.72)	0.379 (3.69)	0.82
Ethanol by direct hydration of ethylene	−2.306 (−5.15)	0.074 (2.68)	0.36
Ethylene oxide by direct oxidation of ethylene	−2.443 (−11.69)	0.140 (14.82)	0.88
Glycerin from acrolein (see note 9)	n.a.	0	n.a.
Isoprene from propylene	−1.125 (−8.07)	0.104 (3.89)	0.75
Melamine from urea	−2.092 (−16.03)	0.173 (7.44)	0.89
Methanol by low-pressure process	−2.035 (−3.28)	0.333 (1.68)	0.59
Phenol from cumene	−1.853 (−19.58)	0.148 (17.36)	0.95
Phthalic anhydride from ortho-xylene	−6.227 (−8.75)	0.234 (7.34)	0.89
Phthalic anhydride—von Heyden process	−3.275 (−12.70)	0.298 (6.00)	0.92
Phthalic anhydride—BASF process	−1.511 (−2.28)	0.322 (1.61)	0.46
Phthalic anhydride—fluid-bed process	−4.684 (−7.38)	0.183 (6.39)	0.91
Titanium dioxide by chlorination	−2.898 (−25.03)	0.177 (16.54)	0.94
Tricloroethylene from ethylene	−3.207 (−8.64)	0.397 (7.21)	0.88
Vinyl acetate by vapor-phase oxidation of ethylene	−0.319 (−4.36)	0.224 (11.91)	0.97
Vinyl chloride by oxychlorination process	−1.117 (−3.22)	0.313 (5.43)	0.86

Source: Simon, *op. cit.*

$$\phi_j = 0.228\ \pi_j + 0.0065\ D_j + 0.082\ A_j \qquad (6.4)$$
$$(2.16) \qquad (3.91) \qquad (1.63)$$
$$(\overline{R}^2 = 0.43)$$

where ϕ_j is the estimate (in Table 6.2) of ϕ for the jth innovation, π_j is an estimate of the average rate of return (measured as the inverse of the pay-out period) of the jth innovation, D_j is the year (less 1935) when the jth innovation was first used commercially, and A_j is a dummy variable that equals 1 if, in the case of the jth innovation, subsequent producers used the innovator's version of the process, and equals zero if subsequent producers decided to invent around the innovator's patent rather than to license it. Since S_j—the investment (measured as the ratio of the size of the investment to the average assets of the relevant firms) required to utilize the jth innovation—and the intercept are not significant, they are omitted in equation (6.4).[10]

The results in equation (6.4) indicate that the coefficients of π_j and D_j have the expected sign and are statistically significant at the .025 probability level. Also, the coefficient of A_j has the expected sign, and is statistically significant at the .05 level. On the other hand, as noted above, the coefficient of S_j is not at all close to being statistically significant. About 40 percent of the observed variation among these innovations in the estimated rates of diffusion can be explained by equation (6.4). This is a much poorer fit than was obtained in previous studies, but the principal reason, according to Simon, is that the available data concerning π_j are poorer than in previous studies.

7. *Technical Leaders and Followers*

What are the characteristics of the firms that are relatively quick to begin using new techniques? What are the characteristics of those that are relatively slow to begin using them? Based on studies of over a dozen major industrial innovations in a number of quite different manufacturing industries, it is reasonably clear that firms where the expected returns from the innovation are greatest tend to be quickest to introduce the innovation and that firms where the expected returns from the innovation are lowest tend to be slowest to introduce the innovation. This, of course, is what we would expect: A firm's introduction of a new technique is delayed if the return is not deemed adequate to offset the risk involved.

Another factor that can influence how quickly a firm begins to use an innova-

10. Of course, equation (6.4) is only a local approximation. For example, one would not expect it to hold very well when π_j is close to zero. Also, the basic data regarding π_j and A_j often are poor. Thus, equation (6.4) should be viewed with great caution and regarded as tentative. Also, note that, both here and in the next chapter, the estimate of π_j is not normalized as it was in earlier studies because of lack of data.

tion is its size. Holding constant the profitability of the innovation, big firms often tend to introduce an innovation before small firms. In some industries, this may be due to the fact that larger firms—although not necessarily the largest ones—tend to be more progressive than smaller firms. But even if the larger firms are not more progressive and do not introduce more than their share of the innovations, one might expect them to be quicker, on the average, to begin using a new technique than smaller firms, for reasons discussed elsewhere.[11] Thus, there is no contradiction between this finding and the conclusion in Chapter 3 that, in most industries, the largest firms do not seem to introduce more innovations, relative to their size, than somewhat smaller firms. Also, there is no contradiction between this finding and the hypothesis in section 5 that the rate of diffusion of innovations does not tend to be higher in more highly concentrated industries.

However, for innovations where the costs of introduction are relatively low in comparison with the assets of the firms in the industry, this effect of firm size is less likely to show up, since (for one thing) this effect arises partly because bigger firms are better able to finance the large investments required for some innovations, and to take the risks involved. In accord with this proposition, there is a significant inverse relationship between firm size and how rapidly a chemical firm adopted the three most costly innovations in Simon's sample, but little or no such relationship for most of his other innovations.[12]

Other factors that often seem to be associated with how rapidly a firm begins using a new technique are the education and age of the relevant management personnel. Both in agriculture and in the tool and die industry, there is evidence that firms with better-educated and younger managers tend to be quicker to begin using new techniques than firms with less educated and older managers. In industries with bigger firms, no such tendency is found, but this may be because the data used pertain to the wrong managers—the presidents of the firms, who are quite far removed from the decision to introduce many innovations.[13]

Before leaving this subject, it is important to note that the firms that are relatively slow to begin using an innovation tend to substitute it for older techniques more rapidly than those that are quick to begin using it. In other words, the late starters tend to "catch up." For example, railroads that began using diesel locomotives in the fifties made the transition from 10 percent of full dieselization to 90 percent of full dieselization much more quickly, on the average, than did railroads that began using diesel locomotives in the forties. As we shall see in Chapter 7, the same sort of phenomenon is found in the case of numerical control

11. See Mansfield, *Industrial Research and Technological Innovation, op. cit.*

12. Simon, *op. cit.*

13. E. Rogers, *Diffusion of Innovations* (New York: Free Press, 1962); Mansfield, *Industrial Research and Technological Innovation, op. cit.;* and Mansfield *et al., Research and Innovation, op cit.*

in the aircraft, electrical equipment, farm machinery, printing press, computer, farm machinery, tool and die, coal-mining equipment, and other industries. It is also worth noting that, holding constant the date of first use of the innovation, small firms tend to be quicker to substitute the new technique for the old than large firms. In part, this is due to the greater homogeneity of conditions and the smaller number of quasi-autonomous decision-making centers in a small firm. In the chemical industry, the intrafirm rate of diffusion is often almost instantaneous, because one process is often substituted entirely for another. In cases where this is not the case, the intrafirm rate of diffusion seems to be inversely related to a firm's share of the market for the relevant product.[14]

8. The Diffusion of Product Innovations

In previous parts of this chapter, we have been concerned almost entirely with process innovations. What factors influence the rate of diffusion of a product innovation? In other words, once a firm introduces a new product, what factors influence how quickly other firms imitate the new product? Several hypotheses seem important in this regard. First, one would expect that, all other things equal, innovations that were very profitable to the innovator would be imitated more quickly than those that are not so profitable. In most cases, the potential profits to be obtained from imitating the innovation would be expected to be directly related to the size of the profits earned from the innovation by the innovator. And the greater the potential profits from imitation, the sooner one would expect it to occur.

Second, one would expect that, all other things equal, imitation would occur more slowly if entry is blocked by patents, and if the patents have a long time to run, than if these conditions do not hold. Clearly, patents can be a major barrier to entry by competitors unless, of course, the patentee is willing to grant licenses to other firms. And the effective life of the patent can be greater or less than seventeen years since patents often are granted after or before the time when the innovation is first introduced.

Third, one would expect that, all other things equal, imitation would occur more rapidly if only a small investment is required to build a plant of reasonably efficient size to produce the new product. The bigger the necessary investment, the greater the difficulty for potential entrants in obtaining the necessary capital. However, if there are many potential entrants that have very large assets (relative to the differences among innovations in required investment), this factor may not be of importance.

Fourth, one would expect that, all other things equal, imitation would occur more rapidly if it is relatively cheap for potential imitators to "invent around"

14. Simon, *op cit.*

whatever patents the innovator holds on the processes used to produce the product. There are, of course, considerable differences among innovations in the difficulty and cost of inventing around such patents. In some cases, it is quite easy and cheap, whereas in other cases, it may be very expensive. It seems likely that imitation will occur more rapidly in the former cases than in the latter ones. However, in cases where there are a number of potential entrants, all of which are very large relative to the differences among innovations in these costs, this factor may not be of importance.

Fifth, one would expect that, all other things equal, imitation would occur more rapidly if the sales of the product innovation grow rapidly rather than slowly. In cases where the market for the innovation is growing rapidly, potential imitators are likely to feel that it is easier to gain a foothold than in cases where there is little growth. However, it is important to note that the line of causation may proceed in both directions. That is, a rapid rate of imitation may promote a more rapid rate of growth of the market (by resulting in price declines and more sales effort). Thus, rapid imitation may influence, as well as be influenced by, the rate of growth of the market for the innovation.

9. Econometric Results for Twenty-nine Chemical-Product Innovations

To test the hypotheses presented in the previous section, Simon collected data concerning the rate of which twenty-nine product innovations in the chemical industry were imitated.[15] Two measures of the rate of imitation were calculated: (1) the number of years that elapsed from the date when the innovation was first introduced to the date when a second firm began to produce the same product, and (2) the number of years that elapsed from the date when the innovation was first introduced to the date when half of the eventual number of firms producing the product (to date) have entered the market. We let X_j be the first measure for the jth innovation, and Y_j be the second measure for the jth innovation. The values of both measures are given in Table 6.3 for each innovation.

Two regressions were run, the first using X_j as the dependent variable and the second using Y_j as the dependent variable. In both cases, the independent variables are U_j, an estimate of the rate of return to the innovator from its investment in the jth innovation (measured as the inverse of the pay-out period); L_j, the number of years after the introduction of the jth innovation that the innovator's patents on the new product had to run; B_j, the estimated average investment (in millions of dollars) to build a plant to produce the jth innovation; R_j, an estimate of the R and D costs that would have been required to invent around the innovator's process patents relating to the jth innovation; and G_j, the annual rate of

15. *Ibid.*

TABLE **6.3**

Values of X_j *and* Y_j, *29 Chemical-Product Innovations*

PRODUCT	X_j	Y_j	PRODUCT	X_j	Y_j
Acetal resins	2	0	Polyester fiber	6	13
Acrylates	3	6	Polyethylene (LD)	0	12
Acrylic fibers	2	4	Polyethylene (HD)	3	4
Acrylonitrile	12	12	Butyl rubber	21	0
Fluorocarbons	22	26	Polypropylene	2	4
Maleic anhydride	1	29	Tolylene diisocyanate	0	7
Methyl methacrylate	1	1	Polyvinyl alcohol	9	9
Caprolactam	6	6	Carboxymethyl cellulose	1	4
Nylon 6 fiber	0	10	Polybutadiene	0	0
Nylon 66 fiber	14	24	Polyisoprene	3	3
Styrene-butadiene rubber	0	0	Ethylene-propylene rubber	2	3
Polycarbonates	0	0	Perchloroethylene	0	20
Neoprene	38	0	Tetraethyl lead	25	25
Dimethyl terephthalate	4	7	Trichloroethylene	8	31
Terephthalic acid	2	2			

Source: Simon, *op. cit.*

growth of output of the *j*th innovation (measured in percent per year) from the fourth to the fourteenth year after its initial introduction. It is important to note that the data concerning R_j, and U_j are very rough.[16]

Using X_j as the dependent variable, Simon found that

$$X_j = 11.179 - \underset{(2.13)}{5.444}\, U_j + \underset{(3.00)}{0.452}\, L_j + \underset{(4.50)}{0.418}\, B_j + \underset{(1.22)}{0.287}\, R_j - \underset{(0.08)}{0.005}\, G_j - \underset{(3.42)}{0.310}\, T_j, \quad (6.5)$$

where T_j is the year when the *j*th innovation was first introduced less 1922. T_j was included in the equation to take account of the fact that there may have been a trend toward more rapid imitation of product innovations in the chemical industry, just as there has been a trend toward more rapid diffusion of process innovations in this industry. (See section 6.)

16. The data regarding X_j, Y_j, B_j, and G_j were obtained by Simon from trade journals, the engineering literature, and other published sources. He obtained the data regarding U_j and L_j partly from published sources and partly from interviews with the relevant firms. The data regarding R_j came largely from interviews, and seem to be the least trustworthy part of the analysis. Since some of these data are extremely rough, the results should be treated with caution, and regarded as tentative. For a detailed description of the derivation of Simon's basic data, see Simon, *op. cit.*

Using Y_j as the dependent variable, the results are

$$(6.6) \quad Y_j = \underset{(3.34)}{16.603} - \underset{(2.48)}{12.486}\, U_j + \underset{(2.59)}{0.553}\, L_j - \underset{(0.73)}{0.145}\, B_j - \underset{(0.45)}{0.162}\, R_j + \underset{(0.02)}{0.003}\, G_j + \underset{(2.27)}{0.935}\, N_j - \underset{(0.43)}{2.566}\, W1_j - \underset{(3.47)}{13.692}\, W2_j - \underset{(3.17)}{16.224}\, W3_j,$$

where N_j is the eventual number of firms that have come to produce the jth innovation (to date), and $W1_j$, $W2_j$, and $W3_j$ are dummy variables that are 1 when the jth innovation is a fiber, a plastic, or an elastomer, respectively, and are zero otherwise. The reason for including N_j in this equation is self-evident: clearly, one would expect it to take longer for half of the eventual number of producers to have entered the market if the eventual number of producers is large than if it is small. The Ws were included to distinguish among various product groups, which (holding other factors equal) might tend to have different average rates of imitation.[17]

The results in equations (6.5) and (6.6) are generally in accord with the hypotheses in the previous section. The coefficient of U_j has the expected sign in both equations, and is significantly different from zero in equation (6.6). The coefficient of L_j has the expected sign, and is statistically significant in both equations. The coefficient of B_j has the expected sign and is statistically significant in equation (6.5), but not in equation (6.6). However, the coefficients of R_j and G_j are statistically significant in neither equation. Thus, the profitability of the innovation, the existence and duration of patents on the product, and the size of the investment required to produce the innovation seem to have the expected effects on the rate of imitation.[18] But there is no evidence that the size of the R and D expenditures required to invent around process patents or the growth rate of the innovation's sales, have an effect on the rate of imitation.

Equation (6.5) can explain about four fifths of the observed variation in X_j, and equation (6.6) can explain about two thirds of the observed variation in Y_j. Thus, the bulk of the observed variation in the rates of imitation can be explained by the profitability of the innovation, the existence and duration of product patents, and the size of the investment required to produce the innovation (together with T_j in equation (6.5) and N_j and the Ws in equation (6.6)). The statistical sig-

17. The Ws were also included in equation (6.5), but proved not to be statistically significant. Also, T_j was included in equation (6.6), but it did not improve the results.

18. The size of the investment seems to have an effect only on the time lag until a second producer appears. As noted above, the fact that the potential entrants are generally very large probably accounts in part for its nonsignificance in equation (6.6).

nificance of T_j in equation (6.5) suggests that there may have been a trend toward more rapid rates of diffusion for product innovations as well as process innovations in the chemical industry. The statistical significance of N_j and some of the *W*s bears out the hypotheses regarding them stated earlier in this section.

10. Public Policy and the Rate of Application of New Technology

Finally, it is important to note that there are many ways that public policy influences the rate of application of new technology—that is, the rate at which new technology is commercialized, and the rate at which the resulting innovations spread. For example, governments often promote the use of new technology by encouraging and financing research and development. This role of government has, of course, been very significant: for example, in the United States the federal government finances over one half of the nation's research and development. The composition of government spending on R and D—the extent to which it is geared to defense or civilian needs, and the extent to which it is directed at one industrial or technological sector rather than another—is clearly an important determinant of which kinds of new technology tend to be introduced particularly rapidly.[19] Recognizing that their scientific and technological resources are limited, many countries have decided to specialize in particular areas cf science and technology. For example, in France, criteria have been established for the support of science and technology on the basis of market and technological opportunities, and it has been suggested that the nation be "active" in certain fields, but only "vigilant" in others.[20]

In addition, the government's procurement policies—both in R and D and elsewhere—can be important in other ways. For example, it has been suggested that performance criteria, which specify the desired end product without limiting the design to existing products, be substituted where possible for product specifications in government procurement. Proponents of performance-based procurement argue that it will free industry to innovate, limited only by the requirement that it perform certain specified functions; encourage cost reduction for the government; and encourage the government to serve as a pilot customer for technical innovations in areas where it represents a big enough market or a market sufficiently free from local restrictions, codes, etc., to make it worth industry's

19. Also, there are various government programs whereby loans are made to industry to promote the application of new technology, as well as tax incentives. Such programs have been adopted by many countries. See. E. Mansfield, "Federal Support for R and D Activities in the Private Sector," in *Priorities and Efficiency in Federal Research and Development* (Washington, D.C.: Joint Economic Committee of Congress, 1976).

20. *The Conditions for Success in Technological Innovation* (Paris: Organization for Economic Cooperation and Development, 1971).

while to innovate.[21] Also, in the R and D area, the way in which the government procures R and D can have important effects. For example, many governments have tended to convert government laboratories and increase the amount of government-financed R and D done in industrial firms in order to bring R and D into closer contact with application and commercialization.[22] In addition, in the United States, the government has made efforts of various kinds to stimulate "spillover" of technology from defense and space projects to the civilian economy. (However, the extent of their success to date is by no means clear.)[23]

Another important factor influenced by government policy is the educational system. It need hardly be pointed out that the rate at which new technology is applied is dependent on the number of scientists and engineers that a nation educates. But this is only part of the story. The general educational level of a nation's work force, not simply the number of scientists and engineers, will surely have an important effect. Moreover, the educational level—and type of education—of a country's managers will have an extremely important effect too. This latter point is emphasized in recent OECD reports and elsewhere, but it cannot be overemphasized. Too often in the past—and at present—the educational system's role in this area has been viewed almost entirely in terms of scientific and technical education, whereas in fact the proper education of managers may be at least as important.[24]

Still another way that governments influence the rate of application of new technology is through their policies regarding industrial competition and monopoly. In Chapter 1, we described some of the effects of market structure on the rate of technological change. Moreover, the government's attitude toward union policies aimed against the application of new technology are important. So, too, are its attitudes toward building codes and other devices that have been used to obstruct the use of new technology. In addition, since they influence the incentives for innovation, the tax laws are of importance. The patent laws too have an effect, but it is doubtful that changes in the patent system of the sort proposed recently in the United States would be of primary importance.[25]

Finally, governments influence how rapidly new technology is applied by their policies regarding full employment, the distribution of the social costs of technological change, and the assessment of the social costs of new technology. Reasonably full employment, coupled with a high rate of investment, tends to

21. E. Mansfield, *The Economics of Technological Change* (New York: Norton, 1968).

22. See *The Conditions for Success in Technological Innovation, op. cit.*

23. For some discussion of the extent and type of such "spillover," see *The Commercial Application of Missile-Space Technology* (Denver: Denver Research Institute, 1963).

24. See *Gaps in Technology: General Report* (Paris: Organization for Economic Cooperation and Development, 1968).

25. See Mansfield, *The Economics of Technological Change, op. cit.*, ch. 7.

promote the application of new technology and the rapid and relatively painless adaptation of the labor force to technological change. Programs that spread more equitably the social costs of technological change—for example, retraining programs that help displaced workers and programs of continuing education that prevent professional obsolescence—reduce resistance to new techniques. The desirability of an effective mechanism for technology assessment, if one can be devised, is clear enough. However, given our very limited capacity for technological forecasting and the tremendous problems of evaluating the various effects of a new technology, it is important that, while exploring the usefulness of various types of technology-assessment programs, we guard against the possibility that these programs are turned into devices for the retardation of needed technological change.[26]

11. Summary and Conclusions

Some years ago, a model was proposed by one of the authors to represent the process by which the use of an innovation spreads. This model has proved useful for forecasting purposes. For example, one of the authors used it in a study carried out for the Small Business Administration at the beginning of 1968 to forecast the percent of firms in the tool and die industry that would be using numerically controlled machine tools in 1970. Its forecasts were better than those resulting from a number of alternative methods. However, the model is by no means complete: it can be—and is being—extended to include a number of additional variables, such as the extent of the industry's R and D expenditures and its market structure. Much more will be said on this score in Chapter 7.

In this chapter, results were reported which indicate that this model can be used to represent the growth over time in the percent of industry output produced with a new process. The basic model can easily be reinterpreted in these terms. Using data regarding twenty-three process innovations in the chemical industry, a study was made which indicates that the logistic curve generally fits quite well, and that the rate of diffusion is directly related to the profitability of the innovation. Also, whether or not the imitators licensed the process from the innovator may have affected the rate of diffusion, and there seemed to be a trend toward more rapid rates of diffusion in recent years.

Turning to the diffusion of product innovations, it was hypothesized that the

26. See U.S. House Committee on Science and Astronautics, *Technology: Process of Assessment and Choice* (Washington, D.C.: National Academy of Sciences, 1969); U.S. House Committee on Science and Astronautics, *A Study of Technology Assessment* (Washington, D.C.: National Academy of Engineering, 1969); and Office of Technology Assessment, *Annual Report to the Congress* (Washington, D.C.: Government Printing Office, 1976).

rate at which a new product will be imitated is dependent on the profitability of producing the innovation, the existence and duration of patents on the innovation, the size of the investment required to produce the innovation, the cost of inventing around the innovator's relevant process patents (if they exist), and the rate of growth of the new product's sales. An econometric study of twenty-nine product innovations in the chemical industry was carried out. The results indicate that the profitability of producing the innovation, the existence and duration of patents on the innovation, and the size of the investment required to produce the innovation have a significant effect on the rate of imitation, but that the other two factors do not. Also, as in the case of process innovations, there is some evidence that, all other things equal, the rate of imitation has tended to increase over time.

There are many ways that public policy influences the rate of application of new technology. Governments often promote the use of new technology by encouraging and financing research and development. The government's procurement policies—both in R and D and elsewhere—can be important in a variety of ways. For example, it has been suggested that performance criteria be substituted, where possible, for product specifications. Also, the government's educational policies have an important effect on the rate of application of new technology. The government's policies regarding monopoly and industrial concentration, its policies toward labor organizations, and its tax and patent policies can also be important. Also, its policies regarding full employment, the distribution of the social costs of technological change, and the assessment of the social costs of new technology can be very important too.

7 THE DIFFUSION OF NUMERICALLY CONTROLLED MACHINE TOOLS IN TEN MANUFACTURING INDUSTRIES

1. Introduction

In the previous chapter, we discussed briefly the diffusion of one of the twentieth century's most important innovations—numerically controlled machine tools—in the tool and die industry. In this chapter, we report a detailed study of the diffusion of this innovation in ten manufacturing industries in the United States. Our purpose is to provide a further test of the model described in Chapter 6. In particular, we are interested in seeing how well this model can explain the increase over time in the percentage of new machine tools purchased that have numerical controls. This is a different measure of the rate of diffusion than the two measures—the increase over time in the percentage of firms using the innovation, and the increase over time in the percentage of output produced with the innovation—investigated in the previous chapter. Needless to say, each of these measures is of considerable importance in its own way.

Also, we look at the effects on the rate of diffusion of an industry's market structure and the extent of its investment in R and D. Although previous studies have touched on these factors, this study goes further in measuring their effects than any past work. Further, we investigate how the characteristics of the early users of numerical control (NC) differed from those that were slower to use it, and we study the determinants of the intrafirm rates of diffusion.

2. *NC Machine Tools: Nature, History, and Rates of Diffusion*

To begin with, we must describe NC machine tools. NC is a method of controlling a machine tool by means of numerical instructions expressed in coded form, usually on cards or tapes. These cards or tapes are put on a control unit which interprets these instructions, and the machine is led through some desired sequence of movements and operations. The development of NC started about 1949. The prime mover behind this development was the United States Air Force, which wanted the aircraft industry to have new machining methods that could produce highly complex parts cheaper, faster, and more accurately than conventional methods, and which provided financial support for the development. Due primarily to the work of John Parsons and of the Servomechanisms Laboratory of MIT, experimental models were soon developed. By 1951, numerically controlled machines were being tested in a few plants, and in 1955 the first commercial models went on display at the National Tool Builders Association show.[1] Since then NC has made rapid inroads into manufacturing; by 1972, NC represented over 21 percent of the value of shipments of all metal-cutting machine tools.

To study the diffusion of NC machine tools, we chose a sample of firms in ten manufacturing industries: aircraft engines, airframes, coal-mining machinery, digital computers, farm machinery, industrial instruments, large steam turbines, machine tools, printing presses, and tools and dies. Questionnaires were sent to the firms, and (due to an intensive follow-up campaign) a response was obtained finally from 140 firms.[2] To measure the rate of diffusion in each industry, we collected data on how rapidly NC machines displaced conventional machines in the industry's purchases of new machine tools. Specifically, data were obtained regarding the percent of new machine tool purchases that were NC machines over three three-year periods. Then, as a preliminary estimate of this rate of displacement, linear interpolation or extrapolation was used to approximate the number of years that elapsed from the date when the industry began

1. There are numerous books and articles on NC. A good place to begin is R. Wilson, *Numerical Control in Manufacturing* (New York: McGraw-Hill, 1963), or issues of the specialized journal *N/C World*.

2. The industries chosen were defined more finely than the four-digit SIC code. They were made up of firms primarily involved in producing the indicated products. Firms producing only allied or related products, although in many cases classified within the same four-digit SIC code, were excluded. The industrial-instruments industry consists of a special part of the industry, the manufacturers of automatic valves and actuators.

using NC to the date when 25 percent of its new machine tool purchases were NC. The results are given in Table 7.1. They suggest that large interindustry differences exist in the rate of diffusion: the estimated number of years ranges from 5.9 to over fifteen years.[3]

TABLE **7.1**

Approximate Number of Years from First Adoption to Date When 25 Percent of Industry's New Machine Tool Purchases Were NC, Ten Industries

INDUSTRY	NUMBER OF YEARS
Aircraft engines	12.7[a]
Airframes	12.8[a]
Printing presses	12.3[a]
Coal-mining machinery	10.3[a]
Digital computers	9.5[a]
Large steam turbines	8.6[a]
Machine tools	9.4[a]
Farm machinery	15.0[c]
Tools and dies	12.7[b]
Industrial instruments	5.9[b]

[a] Calculated by linear interpolation.
[b] Calculated by linear extrapolation.
[c] Since, as of 1968, only about 5 percent of this industry's new machine tool purchases were NC, linear-extrapolation techniques seemed inappropriate. The figure of fifteen years seems a conservative estimate based on the industry's reaction to NC to date.

3. The Basic Model

As you will recall from the previous chapter, our basic model was proposed over a decade ago to explain the rate of imitation. According to this model, the proportion of firms not using an innovation at time t that begin using it by time $t + 1$ is a function of: the proportion of firms that are already using the innovation at time t; the profitability of the innovation; the size of the initial investment required; and other "unspecified variables." Based on these assumptions, we have seen that the growth over time in the proportion of firms using an innovation in an industry corresponds to a logistic curve. Specifically,

3. Note too that there were also large interindustry differences in the length of time before 50 percent of the firms in an industry were using NC and in the rate of increase in the percent of the industry's production that was done on NC.

$$P_{ij}(t) = [1 + exp(-l_{ij} - \phi_{ij}t)]^{-1} \tag{7.1}$$

where $P_{ij}(t)$ is the proportion of firms in industry i using innovation j at time t. Furthermore, the parameter ϕ_{ij} can be interpreted as a measure of the rate of imitation; and it can be shown that the expected value of ϕ_{ij} for an innovation in a particular industry is a function of the profitability of the innovation, the size of investment required, and the "unspecified variables."

It is a simple matter to adapt this model of the imitation process to the displacement process described in the previous section. Really, all that is required is a simple redefinition of terms. Let $\chi_i(t)$ be the proportion of industry i's new machine tool purchases that were NC at time t, and let r_i be the maximum proportion of new machine tool purchases that could eventually become NC. Defining $\delta_i(t) = \frac{\chi_i(t+1) - \chi_i(t)}{r_i - \chi_i(t)}$, we can restate the basic assumptions as follows: $\delta_i(t)$ is a function of $\frac{\chi_i(t)}{r_i}$, the profitability of the innovation, and other "unspecified variables." From these assumptions, it follows that

$$\chi_i(t) = r_i[1 + exp(-q_i - \psi_i t]^{-1} \tag{7.2}$$

where q_i is a constant of integration and ψ_i is a measure of the rate of diffusion.

4. *Additional Variables*

Previous studies based on this model, such as many cited in Chapter 6, have used this model to explain differences in the rate at which a number of innovations have spread in the same industry. In contrast, we are interested here in comparing how rapidly the same innovation spread in a number of industries. The present approach has the advantage that we can study more readily the effects of certain variables, which have received limited attention before, on the rate of diffusion in an industry.

Specifically, one variable on which we focus attention is the industry's market structure. The effects of this variable are the subject of controversy. As pointed out in the previous chapter, some economists seem to believe that less competitive industries will have higher rates of diffusion, while others believe that the opposite is true. To test this proposition, we must somehow quantify the notion of market structure, which, of course, has many dimensions. As in almost all empirical work relating market structure to behavior and performance, we were forced by practical considerations to use as our index of market structure, the degree of seller concentration in an industry.

There are a variety of ways that one can measure the degree of seller concentration.[4] This study has employed a two-dimensional approach, taking as its measures the number of firms in an industry and the variance of the distribution of the natural logarithm of firm sizes (measured in terms of employment).[5] This approach, recommended by Hart and Prais and others, has many advantages. Our own expectation is that the number of firms will be positively related, and the variance of the natural logarithm of firm size will be inversely related, to the rate of diffusion. In other words, we would expect that, holding other factors constant, the innovation would tend to spread more rapidly in less concentrated industries. This would be consistent with the tentative conclusions of our previous studies.[6]

Another variable that is included here, but that has not been included in previous studies, is a scale variable pertaining to the industry. What role does this scale variable play? Other things held constant, industries that operate on a larger scale have to do more in an absolute sense to achieve a certain rate of diffusion than industries operating at lower scales. Holding the number of firms constant, the firms in larger-scale industries have to do more investing in NC to achieve some specific rate of displacement. A given percent of machine tool purchases that are NC machines represents a higher absolute level of NC purchases in larger-scale industries. Thus, the effect of increases in this scale factor may be to slow down the diffusion process.[7]

4. For example, see T. Scitovsky, "Economic Theory and the Measurement of Concentration," in *Business Concentration and Price Policy* (Princeton: National Bureau of Economic Research, 1955); P. Hart and S. Prais, "The Analysis of Business Concentration," *Journal of the Royal Statistical Society* (1956); P. Hart, "On Measuring Business Concentration," *Bulletin of Oxford Institute of Statistics* (1957), and "Statistical Measures of Concentration vs. Concentration Ratios," *Review of Economics and Statistics* (1961); J. Blair, "Statistical Measures of Concentration in Business," *Bulletin of Oxford Institute of Statistics* (1956); S. Prais "A Rejoinder," *Bulletin of Oxford Institute of Statistics* (1957); and M. Adelman, "The Measurement of Industrial Concentration," *Review of Economics and Statistics* (1951).

5. This approach dates back to R. Gibrat, *Les Inégalités économiques* (Paris: 1931), who found that the lognormal distribution was a good approximation to the distribution of various economic variables. Hart and Prais also argue in its favor. If the distribution of firm sizes is lognormal and one knows the variance of the distribution and the number of firms, it is possible to derive the concentration ratios. A problem with this is that the lognormal distribution, if not exactly appropriate, is likely to provide a poor fit around its tail so estimates of (top four) concentration ratios are likely to be poor.

6. See E. Mansfield, *Industrial Research and Technological Innovation* (New York: Norton, for the Cowles Foundation for Research in Economics at Yale University, 1968); and E. Mansfield, J. Rapoport, J. Schnee, S. Wagner, and M. Hamburger, *Research and Innovation in the Modern Corporation* (New York: Norton, 1971).

7. Of course, an industry's capacity to invest may be directly related to this scale variable, which may tend to offset the negative effect on the rate of diffusion via another

Finally, a third variable that is included here, and that has received some attention in Chapter 6, is the industry's R and D expenditures as a percent of sales. In studies of technological change, R and D has generally been considered as an input into the inventive process. However, it may well have an important effect on the diffusion process as well. Perhaps the most obvious such effect comes about when the R and D is done in the area of the new technology. The more such R and D that is done, the greater the familiarity with the innovation; and if the innovation is indeed worthwhile, this greater familiarity should, *ceteris paribus,* result in more rapid acceptance of the innovation. However, even when their R and D is not in the area of the new technology, more R and D–intensive industries may be quicker to adopt new technology because they are more likely to be able to understand a new technology (even one not directly related to their work) and, if it is worthwhile, to be more receptive to it.[8]

5. *Econometric Results*

Our first step was to see whether, in accordance with our model, the logistic curve is a good approximation to the growth over time in the percentage of each industry's new machine tool purchases that were NC. To find out, we fitted equation (7.2) to the data for each industry.[9] The results, shown in Table 7.2, indicate

route. For a given level of seller concentration, firms in larger industries may have greater market power. See W. Shepherd, *Market Power and Economic Welfare* (New York: Random House, 1970), for a discussion of how industry size may affect market power.

8. Also, to the extent that a high ratio of R and D expenditures to sales is associated with more progressive management and other such factors, the expected relationship will be found, but it may not indicate a causal connection.

9. The procedure for fitting equation (7.2) was generally to regress the natural logarithm of $\chi_i(t) \div (1 - \chi_i(t))$ on time. The regressions were based on four observations: the percentages of new machine tool purchases that were NC in 1961–1963, 1964–1966, 1967–1969 (where t was measured as the midpoint of the years) and an assumed percentage of 1 percent in the year of first adoption. The industrial-instruments industry regression was based on only three observations, since adoption did not occur until 1965. Although the number of observations is small, the results would be much the same if time were subdivided more finely, and the number of observations were larger. Thus, the small number of observations is really not a problem.

In addition, for industries that adopted in 1962, or 1965, the period percentages were adjusted to reflect the fact that in one year of the period, NC purchases must have been zero. For example, if an industry adopted in 1962, the figure given for 1961–1963 was multiplied by 1.5, since the percentage in 1961 must have been zero. The resulting figure was taken as the percentage for the midpoint of 1962–1963. That midpoint was given the value 1962.5. A similar procedure was followed for industries that adopted in 1965. In the tool and die industry the maximum potential purchases were taken to be 70 percent.

Of course, it may be unrealistic to assume that the upper asymptote is 1 (and 0.70 in

that this equation is a useful representation of the data.[10] Thus, ψ_i can be interpreted as a measure of the rate of diffusion of NC machine tools in the ith industry. The estimates of ψ_i are provided in Table 7.2. As in Table 7.1, interindustry differences in the rate of diffusion are clear.[11]

How can we explain the interindustry differences? Following the discussion above, we could hypothesize a relationship of the form.[12]

$$\psi_i^* = \alpha_0 \pi_i^{\alpha_1} N_i^{\alpha_2} V_i^{\alpha_3} G_i^{\alpha_4} R_i^{\alpha_5} E_i \quad (7.3)$$

where

$$\alpha_1, \alpha_2, \alpha_5 > 0$$
$$\alpha_3, \alpha_4 < 0$$

and where[13]

ψ_i^* = the estimated value of ψ_i (which measures the rate of diffusion in industry i),

π_i = the average rate of return from the investment in NC machine tools in the ith industry (as estimated by the firms in the industry),

---

tools and dies). But if we assume instead that it is 0.60 (and 0.40 in tools and dies), the resulting estimates of ψ_i are almost perfectly correlated with those in Table 7.2, and the results in section 5 remain valid. Thus, our findings are not sensitive to changes of this sort in this assumption.

10. We also fit the logistic function to the growth in the percentage of firms using NC in the industries. The actual and estimated growth patterns were very close, suggesting that our model can describe the imitation (as well as the displacement) process for NC quite well. See A. Romeo, "The Rate of Imitation of a Capital-Embodied Process Innovation," *Economica,* February 1977; and S. Globerman, "Technological Diffusion in the Canadian Tool and Die Industry," *Review of Economics and Statistics,* November, 1975.

11. The values of ψ_i can be related to more intuitive notions of the rates of diffusion. For example, the length of time between the dates when any two percentages of purchases were NC, say F_1 and F_2 (where $F_1 < F_2$), is equal to $\psi_i^{-1} \ln [(100-F_1) F_2 / F_2 (100-F_2)]$. Such lengths of time depend only on ψ_i and not on the constant term q_i.

12. Two variations from the precise form suggested earlier should be noted. First, our basic model suggests a linear form. The multiplicative form is used here because it seemed to fit the data better. Certainly, this form is in keeping with the substance of our model. Second, the size of investment required to introduce NC (as a percent of the average assets of the firm) is not included here, because our data pertain only to the first machine purchased and may be unrelated to the size of subsequent purchases in the replacement process.

13. Average profitability was measured as the average percentage rate of return on NC in an industry, the rate of return being estimated as one hundred times the reciprocal of the pay-out period estimated after adoption by the firms. For a justification of the use of the reciprocal of the pay-out period, see M. Gordon, "The Payoff Period and the Rate of

TABLE 7.2

Estimates of ψ_i and $\overline{R}^2$, Ten Industries

INDUSTRY	ψ_i*	$\overline{R}^2$ (ADJUSTED FOR DEGREES OF FREEDOM)
Aircraft engines	0.264	0.93
Airframes	0.266	0.99
Printing presses	0.283	0.99
Coal-mining machinery	0.328	0.92
Digital computers	0.354	0.97
Large steam turbines	0.403	0.99
Machine tools	0.343	0.83
Farm machinery	0.193	0.63
Tools and dies	0.417	0.68
Industrial instruments	0.643	—[a]

Source: See section 5.

[a] Since there were so few observations for industrial instruments, it seemed misleading to present the value of $\overline{R}^2$. All that one can say about this industry is that the estimate of ψ_i seems to fit the data reasonably well.

N_i = the number of firms in industry i in the United States in 1960,
V_i = the variance of the distribution of the natural logarithm of employment (a measure of the inequality of firm sizes) in industry i in 1960,
G_i = the total value of purchases of all new machinery in industry i in 1960 (a scale variable),
R_i = the average R and D expenditures as a percent of sales for firms in industry i in 1960,
E_i = random error term.

Taking natural logarithms, we can estimate the equation by ordinary least squares. The results are:

Profit,'' *Journal of Business* (1955). N_i is an estimate of the total number of firms in the industry in the United States. To be completely appropriate, the data should reflect geographic markets. In all but three of the industries—printing presses, industrial instruments, and tools and dies—most of the firms compete nationally, and even in these three industries there is some national activity. In any case, there should be a strong correlation between the number of firms nationally and the number in the relevant market. V_i was calculated from the data for the sample firms. G_i was measured as total purchases of new machinery in the industry based on the sample data and roughly corresponding data from the Census of Manufactures. R_i is the average R and D expenditure as a percent of sales in the ith industry, where sales data were gathered from *Moody's*. The data on R and D expenditures were obtained from the firms as answers to a part of the questionnaire.

$$(7.4)\quad Ln\ \psi_i^* = \underset{(4.512)}{0.864}\ In\ \pi_i + \underset{(4.130)}{0.140}\ In\ N_i - \underset{(2.604)}{0.111}\ In\ V_i - \underset{(4.728)}{0.327}\ In\ G_i + \underset{(3.413)}{0.080}\ In\ R_i - \underset{(5.765)}{3.324}$$

$$\bar{R}^2 = 0.742$$

The t statistics are shown in parentheses.

The results seem very encouraging. Every one of the coefficients has the expected sign, and every one is statistically significant at the 5-percent level. Even after adjustment for the small number of degrees of freedom, the independent variables explain over 74 percent of the variation in $In\ \psi_i^*$. Thus, it appears that our model, which is adapted for the first time to explain the rate at which an innovation displaces an older technology with respect to new purchases, can perform this task extremely well. Also, the additional variables on which we focused attention—N_i, V_i, G_i, and R_i—seem quite important to the diffusion process, and certainly worthy of further study.[14]

6. *Individual Firms: Characteristics of Users and Nonusers*

In preceding sections of this chapter, we have compared the behavior of various industries. Let's look now at the response of individual firms in the same industry to the innovation. What factors distinguish users from nonusers of the

14. As noted above, we also gathered data on the imitation process, as well as rough data on the rate at which NC had replaced conventional methods in the production processes of the firm. We thus were able to measure the rate of imitation, and a production-based version of the rate of replacement. The three measures of the rate of diffusion were only slightly correlated, but we found that a form of equation (7.3) worked quite well when these other two rates were substituted for the rate of replacement. Specifically, coefficients of the relevant variables were of the expected signs and in most cases were statistically significant. Thus, despite the small number of degrees of freedom, we have added confirmation of the explanatory variables' importance.

Another noteworthy point concerns the effect of the R and D variable. Since R and D expenditure as a percent of sales in an industry is correlated with the two concentration variables (the number and size distribution of firms), one might argue that its effect reflects positive effects of concentration of the rate of diffusion. However, when R and D is omitted from the equation, the coefficients of the two concentration variables remain of the same sign, although the coefficient of the variance in firm size is insignificant. Thus, whether or not the R and D variable is included, there is no evidence that increases in concentration result in higher rates of diffusion.

innovation? How do the characteristics of early adopters differ from those of late adopters?

To begin with, we see whether it is possible to distinguish between firms that had begun using NC by 1970 and those that had not, on the basis of the characteristics of the firms. On a priori grounds, several characteristics may be important determinants of the likelihood of use of the innovation. First of all, as noted in Chapter 6, firm size may be important. Larger firms would seem more likely to be using the innovation for three reasons. One, they will tend to have more equipment than smaller firms and, consequently, at any time are more likely to have equipment in need of replacement. Two, larger firms experience a wider range of operations and are more likely to be involved in operations suitable to the use of the innovation. Three, larger firms have more resources available to them and are more likely to be able to finance the investment and to absorb a loss should a risky investment fail.[15]

A second factor that is likely to influence whether or not a firm will use the innovation is the profitability of the innovation to the firm. Other things equal, the higher the expected returns from an investment, the more quickly a firm would be expected to begin using an innovation (and the more likely it is to be using the innovation as of some date).

A third factor affecting a firm's quickness of reaction to an innovation is the progressiveness of its management. Measurements of progressiveness are, of course, difficult to obtain. As a proxy, one might consider the age and education of the firm's president.[16] Younger and better-educated managers might be expected to be less conservative, less tied to old ways of doing things, and less risk-averse. In addition, the better-educated managers may be better able to understand the new technology and, thus, may be more likely to react positively to it.

A useful form of the relevant hypothesis can be expressed as follows:

15. The work of Mansfield, *Industrial Research, op. cit.,* and Mansfield *et al., Research and Innovation, op. cit.,* support this contention. Nabseth and Ray found large firms quicker to adopt in some cases and slower in others, but they did not hold other variables constant. See L. Nabseth, "The Diffusion of Innovations in Swedish Industry," in *Science and Technology in Economic Growth* (New York: Macmillan, 1973), and G. Ray, "The Diffusion of New Technology," *National Institute Economic Review* (1969). This relationship is more likely to show up for costly innovations. See Chapter 6.

16. Considering just the firm's president rather than all members of management involved in the decision-making process assumes that he is the key decision maker or that he sets the tone for the others. While this may not be true, especially in the large firms, the assumption was necessitated by the practicalities of data collection. Later we consider variables pertaining to a wider group of managers in the tool and die industry. Mansfield found that, when included separately, both age and education of the company president were significant determinants of use in the tool and die industry, but that, when included together, only education was significant. See Mansfield *et al., Research and Innovation, op. cit.*

(7.5) $$e^{x_i} = \alpha_0 H_i^{\alpha_1} U_i^{\alpha_2} A_i^{\alpha_3} E_i^{\alpha_4} Z_i,$$

where x_i equals 1 if the ith firm used NC at the beginning of 1970 and zero otherwise; H_i is the firm's size, measured by the number (in thousands) of employees in 1960; U_i is the profitability of the innovation to the firm; A_i is the age of the firm's president (in 1960); E_i is the president's education (in years beyond the eighth grade) as of 1960; α_1, α_2, α_4 are positive; α_3 is negative; α_0 is a constant term which may vary from industry to industry; and Z_i is a random error term.

After taking natural logarithms, we can estimate the αs by least-squares. Since profitability data were available only for users of the innovation, U_i could not be included in this analysis. Based on the remaining variables, the statistical results were as follows (t statistics in parentheses):[17]

(7.6)
$$x_i = \underset{(2.437)}{1.969} + \underset{(7.060)}{0.117} \ln H_i - \underset{(-2.021)}{0.385} \ln A_i + \underset{(1.584)}{0.163} \ln E_i$$
$$\bar{R}^2 = 0.464$$

The results are in accord with our hypotheses. The estimates of α_0, α_1, and α_3 are significant at the 5-percent level, and the estimate of α_4 is significant at the 10-percent level. We allowed the constant term to vary across industries, but an F test indicated that allowing for such interindustry variation did not significantly improve the explanatory value of the equation.[18]

In this type of model, the predicted value of x_i is sometimes interpreted as a conditional probability of use of NC.[19] If we adopt this interpretation, the results suggest that a 10-percent increase in firm size leads, *ceteris paribus,* to a 0.011 increase in the probability of use of NC. A 10-percent increase in age leads, *ceteris paribus,* to a 0.04 decrease in the probability of use. And an increase of 10 percent in the years of schooling of the company's president, other things equal, results in a 0.02 increase in the probability of use of NC.

17. These results were based on the 104 firms for which data on size, age, and education were available. The other thirty-six firms did not provide age and/or education information. Given the rough nature of the data, more involved techniques did not seem worthwhile. See A. Goldberger, *Econometric Theory* (New York: Wiley, 1964).

18. The F test involved testing the hypothesis that the constant terms were all equal. The hypothesis could not be rejected at the 5-percent level.

19. See Goldberger, *op. cit.*, for a discussion of this interpretation of the regression results. One problem with this interpretation is that x_i could be greater than 1 or less than zero. This result, of course, would not be consistent with the probability interpretation.

7. Users and Nonusers in the Tool and Die Industry

How rapidly a firm adopts an innovation is also likely to depend on how rapidly it learns of the innovation and the nature of its decision-making process. From interviews with fifteen firms in the tool and die industry, we gathered data concerning both of these variables.[20] The first variable obtained for each firm was the date of first knowledge of the innovation. This date, the earliest possible time at which a firm could adopt an innovation, may be important in at least two respects. First, the firms to hear first may have heard earlier precisely because they are the ones more interested in new ideas and, thus, more interested in seeking out and applying innovations. Second, as of any given date, the first firms to hear of an innovation would have had more time to digest any information about it and would also have had a longer opportunity to adopt it.

Our other variable is concerned with the organizational structure of the firm. One might argue that more complex organizational structures are not conducive to rapid adoption of an innovation, since the more levels of approval that are required for adoption, the longer such approval takes and the greater the possibility that adoption is held up or prevented by a negative decision at any one point.

To incorporate these variables into the type of discriminant model suggested above, we hypothesize a relationship of the form:[21]

$$x_i = a_0 + a_1 Y_i + a_2 n_i + a_3 H_i + a_4 A_i + a_5 E_i + u_i, \tag{7.7}$$

where x_i, H_i, A_i, and E_i are as previously defined, with A_i and E_i being averages for all managers involved in the decision to adopt (or not adopt) NC; Y_i being the number of years that the managers of firm i have known of NC (and their firm has been in business), as of the beginning of 1970; and n_i being the number of people in the firm who had to approve a decision to adopt NC, in 1960;[22] a_1, a_3, a_5 being positive; and a_2, a_4 being negative; and u_i being an error term.

A linear regression yields the following results:

20. This sample of fifteen was chosen so that it contains roughly the same percentage of users of NC as did the larger sample used in the rest of the study. Specifically, from the list of NC users, six were randomly selected; and from the list of nonusers, nine were randomly selected.

21. This linear form seemed to fit the data for the tool and die industry better than the logarithmic form. The results are qualitatively the same whether a linear or multiplicative form is used.

22. n_i was measured by asking the interviewees what firm members had to approve the decision to adopt or not adopt NC. (Beyond this, the definition of decision maker was left up to the individual firms.) n_i is the number of such members, while A_i and E_i are unweighted averages for these n_i decision makers. For those firms not in existence in 1960, n_i, H_i, A_i, and E_i are measured as of the first full year of operation.

$$(7.8)\quad x_i = 0.032\,Y_i - 0.123\,n_i + 0.014\,H_i - 0.027\,A_i + 0.212\,E_i + 0.129$$
$$(1.257)\quad (1.497)\quad (3.608)\quad (-2.201)\quad (3.135)\quad (0.206)$$

$$\overline{R}^2 = 0.638$$

The coefficients of H_i, A_i, and E_i are all significant at the 5-percent level, the coefficient of n_i is significant at the 10-percent level, and the coefficient of Y_i is significant at the 15-percent level. The t statistics are shown in parentheses below the regression coefficients.

8. *Length of Time before First Use*

The discriminant function in equation (7.6) is, of course, a function of time: as years pass, more and more nonusers will pass into the users group. Confining our attention now to firms that have begun using NC, how do they differ in the length of time they waited before adopting this innovation? Table 7.3 provides, for each industry, the average and standard deviation of the length of time, from the date of first adoption in the industry, that the sample firms (using NC) waited

TABLE 7.3

Average and Standard Deviation of Number of Years that Firms Waited before Beginning to Use NC (as Measured from the Date When NC Was First Used in the Industry)[a]

INDUSTRY	AVERAGE	STANDARD DEVIATION
Aircraft engines	4.00	2.71
Airframes	7.89	4.64
Printing presses	7.29	4.54
Coal-mining machinery	5.00	3.46
Digital computers	6.73	3.84
Large steam turbines	0.50	0.50
Machine tools	3.79	2.23
Farm machinery	3.71	3.09
Tools and dies	4.58	2.61
Industrial instruments	1.17	1.64

Source: See text.

[a] Based on 101 sample firms using NC as of 1970: four in aircraft engines, nineteen in airframes, seven in printing presses, four in coal-mining machinery, fifteen in digital computers, two in large steam turbines, nineteen in machine tools, seven in farm machinery, twelve in tools and dies, twelve in industrial instruments.

before beginning to use it. For a particular firm, what factors determine this length of time? It seems reasonable to believe that the same types of variables that were important in distinguishing users from nonusers will also be important here.[23] In other words, we might hypothesize a relationship of the form:

$$d_i = B_0 H_i^{B_1} U_i^{B_2} A_i^{B_3} E_i^{B_4} u_i \tag{7.9}$$

where d_i is the number of years the ith firm waited before adopting NC (measured from the year the first firm in its industry adopted NC); H_i, U_i, A_i, and E_i are as previously defined; u_i is a random error term; B_1, B_2, B_4 are negative; B_3 is poistive; and B_0 is a constant term that may vary across industries.

Taking natural logarithms and applying least-squares, one obtains the following results (t statistics in parentheses):[24]

$$\begin{aligned} \ln d_i = & -0.673 \ln H_i - 0.426 \ln U_i + 3.408\,[AF] - 0.749\,[AE] \\ & \quad (-3.198) \qquad\quad (-1.025) \qquad\quad (1.889) \qquad\quad (-0.318) \\ & -0.928\,[II] + 3.433\,[CM] + 3.073\,[DC] + 3.172\,[FM] \\ & \quad (-0.555) \qquad (1.676) \qquad\quad (1.743) \qquad\quad (1.933) \\ & +2.708\,[MT] + 1.126\,[PP] + 0.351\,[LT] - 0.128\,[TD] \\ & \quad (1.783) \qquad\quad (0.748) \qquad\quad (0.172) \qquad\quad (0.078) \\ & \bar{R}^2 = 0.31 \end{aligned} \tag{7.10}$$

The last ten variables are dummy constant terms, equal to 1 if the firm belongs to the industry indicated in brackets and zero otherwise. The letters in brackets refer to the ten industries. (AF is airframes; AE is aircraft engines; II is industrial instruments; and so on.)

The results suggest that, *ceteris paribus,* the length of time a firm waits before first adoption will be shorter the larger the firm. The estimate of B_1 is statistically significant. Also, there is a negative, although not quite significant, influence of the profitability of the innovation on the length of this delay. The effects of age and education were not at all close to significant. This equation was also tested holding the constant term constant over the industries, but an F test in-

23. We also gathered data on each firm's growth rate, but it was never significant in explaining any aspect of speed of response to NC. Mansfield found the same thing in previous studies.

24. The results are based on the seventy-three users that provided size and profitability data. Three aircraft firms that adopted NC in 1951 and 1952 were omitted from the sample, because they used machines that were provided by the government. The lag for the other aircraft firms was measured from 1956.

dicated that allowing for interindustry differences added significantly to the explanatory power of the equation.[25]

We can consider the quantitative impacts of size and profitability by looking at the effect on d_i of a 10-percent increase in H_i or a 10-percent increase in U_i. Increasing size of firm by 10 percent, other things equal, leads to about a 6-percent decrease in the lag. Measuring from the mean lag of 4.9 years, this means that a firm will begin using NC about 0.3 years earlier. Increasing the rate of return by 10 percent leads, *ceteris paribus,* to a 4-percent decrease in the lag. Measuring from the mean lag, this means that a firm will begin using the innovation 0.2 years earlier. It is also interesting to compare these results with those obtained earlier by one of the authors. Testing a similar equation, he found a coefficient of firm size of -0.40, which is of the same sign and not significantly different from the coefficient of -0.67 found in this study. His coefficient of profitability ranged from -0.03 to -1.53; the coefficient here (-0.42) falls within that range.

9. *Intrafirm Rates of Diffusion*

The results in the previous three sections were concerned only with how quickly firms begin to use NC. They did not consider how quickly, once the innovation was first adopted, its use spread within the firm. There are various methods of measuring this rate of intrafirm diffusion. In line with the discussion at the industry level, we will consider the rate at which numerical control displaced conventional machine tools in the firm's purchases of new machine tools. An application of our model at the firm level would suggest that the percentages of new machine tool purchases that were NC machines would increase over time according to a logistic pattern, or:

$$C_i(t) = h_i[1 + exp(-a_i - w_i t)]^{-1} \tag{7.11}$$

where $C_i(t)$ is the percent of firm i's new machine tool purchases that were NC machines at time t, and h_i is the maximum possible percent of new machine tool purchases that were NC. The parameters, a_i and w_i, can be estimated by least-squares. As we know, w_i is a measure of the rate of intrafirm diffusion.[26] Table 7.4 summarizes the distribution of the intrafirm rates of diffusion for the firms in

25. The F test rejected at the 5-percent level the hypothesis that the constant terms were all equal.

26. The estimation procedure was carried out at the firm level in the same way as at the industry level. Estimates of w_i were possible for fifty-nine firms; the average value of w_i^* was 0.590. In this case, the h_i were assumed to be 1, but judging from the results described in note 9, the results would be much the same if 0.6 were used instead.

TABLE 7.4

Average and Standard Deviation of Estimated Number of Years from the Time When 10 Percent of a Firm's New Machine Tool Purchases Were NC Machines to the Time When 40 Percent Were NC, by Industry[a]

INDUSTRY	AVERAGE	STANDARD DEVIATION
Aircraft engines	6.1	1.1
Airframes	4.3	0.8
Printing presses	3.1	1.9
Coal-mining machinery	2.8	1.2
Digital computers	3.9	2.5
Large steam turbines	4.8	0.4
Machine tools	3.6	1.4
Farm machinery	6.5	2.5
Tools and dies	3.5	1.0
Industrial instruments	2.3	0.7

Source: See section 1.

[a] Based on a total of fifty-nine observations: three in aircraft engines, twelve in airframes, three in printing presses, three in coal-mining machinery, nine in digital computers, two in large steam turbines, seventeen in machine tools, three in farm machinery, two in tools and dies, and five in industrial instruments.

each industry. For ease of interpretation, the estimates of w_i have been converted to estimates of the number of years between the time when 10 percent of all machine tool purchases were NC to when 40 percent were NC.[27]

The previous sections have suggested some variables that are likely to affect these rates. Size of firm, profitability of the innovation, and perhaps age and education of the company president should be considered. However, when considering the intrafirm rate of diffusion, we might expect a negative effect of firm size. The reasons for this have been discussed (at least briefly) in the last chapter. In addition, the year of first adoption by the firm should also be a determinant of the intrafirm rate of diffusion.[28] As we noted in the previous chapter, the later the year of first adoption, the quicker, *ceteris paribus,* will be the intrafirm diffusion in terms of purchases. This would be expected since the degree to which a firm invests in an innovation will depend on the risks involved. The later the date of first adoption, the less the risks are at that time, since risk can be expected to decrease over time as more experience with the innovation is accumulated outside the firm. A useful form of the relevant hypotheses can be expressed as:

27. The length of time equals $w^{*-1} \ln 6.0$. See note 11.

28. In his study of intrafirm diffusion, Mansfield found size of firm and date of first adoption to have these suggested effects.

$$w_i^* = b_0 H_i^{b_1} U_i^{b_2} A_i^{b_3} E_i^{b_4} F_i^{b_5} \xi_i, \tag{7.12}$$

where w_i^* is the estimated rate of intrafirm diffusion; H_i, U_i, A_i, E_i are as previously defined; F_i is the year firm i first began using NC, measured from 1950; ξ_i is a random error term; b_1, b_2, b_4, b_5 are positive; b_3 is negative; and b_0 is a constant term that may vary from industry to industry.

This equation can be linearized and estimated by least-squares. The results are as follows (t statistics in parentheses):

$$\ln w_i^* = -0.136 \ln H_i + 0.024 \ln U_i + 0.343 \ln F_i - 1.377 \tag{7.13}$$
$$\quad (-2.958) \qquad (0.266) \qquad (2.352) \qquad (-3.483)$$

$$\bar{R}^2 = 0.416$$

The coefficient of firm size is negative and highly significant. The coefficient of profitability is positive, but is not significant; it remains in the equation because both theory and previous empirical results demand its retention. The coefficient of year of first adoption is highly significant. The coefficients of the age and education variables were far from significant and were excluded from the equation. The equation was also estimated allowing the constant term to vary across industries, but this form did not provide any statistically significant additional explanatory power.[29]

The quantitative effects on the intrafirm rate of diffusion of changes in the variables can be easily determined. A 10-percent increase in firm size leads, *ceteris paribus,* to a 1-percent decrease in w_i^*; a 10-percent increase in the rate of return, *ceteris paribus,* leads to a 0.2-percent increase in w_i^*; and a 10-percent increase in the year of first adoption (measured from the mean, this represents initial adoption about 1.2 years later) leads, *ceteris paribus,* to a 3-percent increase in w_i^*.

10. Summary and Conclusions

Based on data obtained from a sample of 140 firms in ten industries, we have tested a number of propositions concerning the diffusion of numerically controlled machine tools. The results indicate that the basic model described in the previous chapter and proposed over a decade ago can explain interindustry differences in the rate of growth of the percentage of new machine tool purchases that are NC. In addition, the results shed new light on a controversial issue: the relationship between market structure and the rate of technological change. In

29. The F test could not, at the 5-percent level, reject the hypothesis that the constant terms were all equal.

the case of numerically controlled machine tools, it is clear that the innovation spread more rapidly, other factors held constant, in less concentrated industries. Thus, with regard to the diffusion stage of the process of technical change, the oft-cited arguments in defense of higher concentration on grounds of "dynamic efficiency" do not seem to be justified in the case of this innovation.

Further, we have shed new light on the role of R and D. Based on our results, R and D has a significant positive effect on the rate of diffusion of an innovation, even when the innovation is not directly related to the areas in which the R and D is done. This finding seems eminently reasonable. Besides producing inventive output, R and D expenditures also result in a firm's having more technically competent personnel who are likely to be better able to evaluate new technologies of any sort. Also, it may be that firms that invest heavily in R and D tend to be less resistant to change.

In addition, our results provide further information concerning the effects of various characteristics of a firm on how rapidly it begins using an innovation and on the intrafirm rate of diffusion of an innovation. In general, the results are quite consistent with the findings of earlier studies. However, they also suggest the importance of a new variable: the number of people having to approve a decision to introduce the innovation. Although significant only at the 10-percent level, this variable seems worthy of further study.

8 SOCIAL AND PRIVATE RATES OF RETURN FROM INDUSTRIAL INNOVATIONS*

1. Introduction

The studies described thus far in this book have been concerned with the processes by which new industrial technology is produced and applied, but not with the returns from investments in new technology. For many years, economic analysts and policy makers have been interested in developing and obtaining better and more complete data concerning social and private rates of return from industrial innovations. It has long been recognized that information of this sort is essential if public policy concerning civilian technology is to be formulated rationally. Yet despite the work of Denison, Fellner, Griliches, Mansfield, Minasian, Peterson, Terleckyj, and others, existing knowledge in this area has been far too weak to provide a reasonably adequate foundation for analysis and policy.[1]

The purpose of this chapter is to report the results of seventeen case studies, each of which estimates the social and private rates of return from the investment

* George Beardsley is a co-author of this chapter. He was responsible in particular for the work described in sections 9 and 10.

1. For a summary of existing knowledge in this area, see E. Mansfield, "Contribution of R and D to Economic Growth in the United States," *Science* (February 4, 1972); and his testimony in *Federal Research and Development Expenditures and the National Economy,* Hearings before the House Committee on Science and Technology, 1976.

in a particular industrial innovation. Also, a detailed account is given of the social and private returns from the investment in innovation by one of the nation's largest firms. The results, although subject to obvious limitations and shortcomings, should provide us with a better understanding of this topic. The studies described here seem to be the first attempts to make direct measurements of this sort in the industrial sector.

2. The Sample of Innovations

Our first step in carrying out this investigation was to contact a number of business firms in the Northeast and to try to persuade them to provide us with data bearing on the social and private returns from innovations that they had carried out. As would be expected, a substantial percentage of those that were contacted refused to cooperate because, despite our assurance that the data would be held in strictest confidence, they felt that such data were too sensitive to show outsiders. Those firms that were willing to cooperate were asked to pick one or more of their recent innovations more or less at random. Then many man-weeks were spent gathering data concerning each innovation and its effects from the innovating firm, from firms using the innovation (if it was used by firms), and from other sources. These innovations occurred in a wide variety of industries (described below), and in firms of quite different sizes. Most of them are of average or routine importance, not major breakthroughs. Although the sample cannot be regarded as randomly selected, there is no obvious indication that it is biased toward very profitable innovations (socially or privately) or relatively unprofitable ones.

To be more specific, the innovations in our sample are as follows:

1. a product innovation in the primary metals industry in the late 1950s that reduced the cost of making household appliances;
2. a product innovation in the machine tool industry in the early 1970s that incorporated a new computer control system;
3. a new type of component for control systems that was introduced in the late 1960s;
4. a new type of construction material introduced in the mid-1950s that reduced substantially the cost of performing certain building operations;
5. a new type of material used in drilling wells, this material having been introduced in the mid-1960s;
6. a new type of drafting for certain types of industrial heating equipment, this being a process innovation introduced in about 1960;
7. a product innovation in paper products introduced in about 1967, which reduced the production cost of the innovator's customers;
8. a new type of thread introduced in the early 1960s that allowed garment manufacturers to lower their production costs;

9. a product innovation in control mechanisms for doors that was introduced in the early 1970s;
10. a new type of electronic device introduced in the early 1970s that reduced the costs of carrying out certain kinds of operations with video tape;
11. a product innovation in the chemical industry (which, here and below, is defined to include petroleum refining) in the late 1960s that reduced costs of users;
12. a process innovation in the chemical industry introduced in the mid-1960s;
13. a process innovation in the chemical industry that reduced the cost of certain aromatic chemicals, introduced in the late 1960s;
14. a major chemical-process innovation introduced in the early 1960s;
15. a product innovation in household cleansing devices introduced in about 1960 that reduced the cost of cleaning floors;
16. a product innovation in stain removers introduced in the mid-1960s;
17. a product innovation in household dishwashing liquids in the early 1960s that reduced costs of operating dishwashers.

Note several things at the outset concerning this sample. First, thirteen of our seventeen innovations are (from the point of view of the innovating firm) product innovations, not process innovations. Second, in the case of ten of these thirteen product innovations, the new product is used by firms, not consumers, and its principal effect is to lower the costs of the using firms. Third, in the case of the remaining three product innovations, the new product is used by consumers. However, these new products did not enable the consumer to do entirely new things; instead, they were meant to reduce the cost of carrying out certain household activities that were already being performed in somewhat different ways.

3. Estimation of Social Benefits: Product Innovations Used by Firms

As noted in the previous section, the innovations in our sample can be divided into three classes: product innovations used by firms, product innovations used by households, and process innovations. Based on an intensive study of each of the innovations in each of these classes, it appears that the same general kind of model is applicable to all of the innovations in our sample in a particular class. In this and the following sections, we describe the model that is used to measure the social benefits in a particular period from product innovations used by firms.

In each case, these new products resulted in a potential saving to users. For example, the product innovation in the primary metals industry resulted in a potential saving to makers of household appliances. Thus, each of these innovations could shift downward the supply curve of the industry using the innovation.

How far downward this supply curve will shift depends, of course, on the pricing policy of the innovator. If the innovator charges a relatively high price for the new product, the supply curve may shift only slightly. Indeed, if the innovator charges a high enough price, the supply curve will not shift downward at all.

Suppose that the innovator decides to set a price for its new product which yields a profit[2] to the innovator equivalent to r dollars per unit of output of the industry using the innovation (for example, r dollars per appliance in the case of the new type of metal). Also, suppose that the industry using the innovation is competitive and that its supply curve is horizontal in the relevant range. In particular, assume that, before the advent of the innovation, this supply curve was S_1 in Figure 8.1, and the price charged by the industry using the innovation was P_1. After the advent of the innovation, this supply curve is S_2, and the price is P_2.

Under these circumstances, the social benefits from the innovation can be measured by the sum of the two shaded areas in Figure 8.1. The top shaded area is the consumers' surplus due to the lower price (P_2 rather than P_1), resulting from the use of the innovation.[3] In addition, there is a resource saving, and a corresponding increase in output elsewhere in the economy, due to the fact that the resource costs of producing the good using the innovation—including the resource costs of producing the innovation—are less than $P_2 Q_2$. Instead, they are P_2Q_2 minus the profits of the innovator from the innovation, the latter being merely a transfer from the producers of the good using the innovation to the innovator. Thus, besides the consumer's surplus arising from the price reduction, there is a resource saving amounting to the profits of the innovator.[4]

However, two adjustments must frequently be made in the estimate corresponding to the lower shaded area in Figure 8.1. First, if the innovation replaces another product, the resource saving cited in the previous paragraph does not equal the profits of the innovator (from the innovation), but these profits less those that would have been made (by the innovator and/or other firms) if the innovation had not occurred and the displaced product had been used instead. This is the correct measure of the resource saving. Second, if other firms imitate the innovator and begin selling the innovation to the industry that uses it, their profits from the sale of the innovation must be added to those of the innovator to get a full measure of the extent of the resource saving due to the innovation.

2. Profit here is defined to be gross of any depreciation on the investment in the innovation. Basically, it is a cash-flow concept.

3. Of course, one problem in any model of this sort is that the concept of consumers' surplus is a controversial one. But as so many writers have pointed out, it is hard to do applied work in this area without adopting some such concept. See, for example, J. Currie *et al.*, "The Concept of Economic Surplus and Its Use in Economic Analysis," *Economic Journal* (1971).

4. To put matters somewhat differently, the sum of the two shaded areas in Figure 8.1 can easily be shown to equal the social value of the extra output of the product (as measured by the area under the demand curve) plus the value of the resources saved.

FIGURE **8.1**

Social Benefit from Product Innovation that Reduces the Costs of the Industry Using It

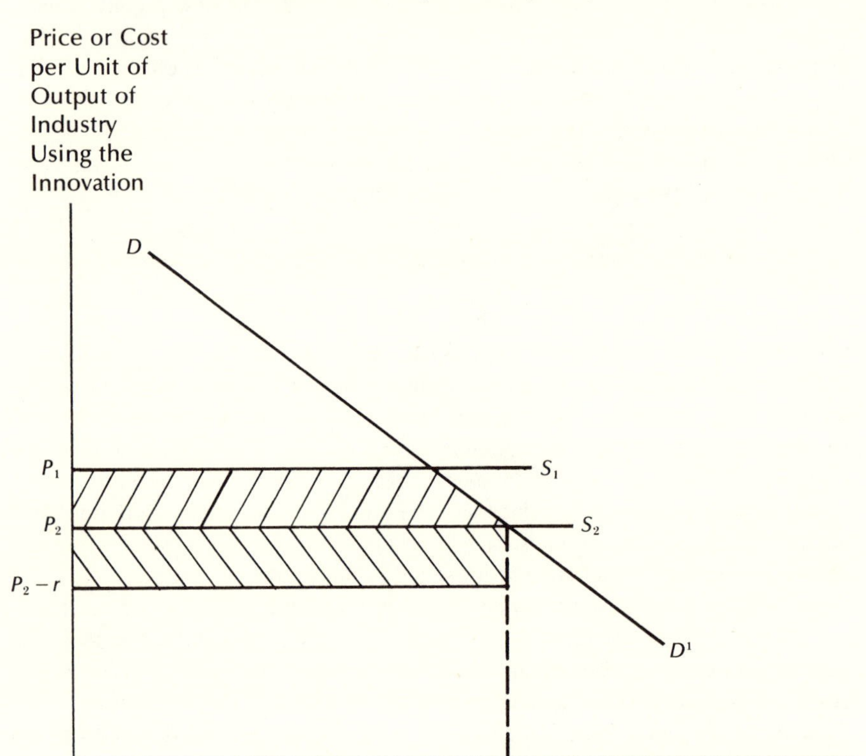

Based on this model, an estimate was made of the social benefit in each period from the investment in each of these innovations. For each innovation, the top shaded area in Figure 8.1 equals

$$(P_1 - P_2)\, Q_2\, (1 - 1/2\, Kn), \qquad (8.1)$$

where $K = (P_1 - P_2)/P_2$ and n is the price elasticity of demand (in absolute value) of the product of the industry using the innovation. To estimate $P_1 - P_2$, we obtained as much information as we could concerning the size of the unit cost reduction due to the innovation in the industry using the innovation. Based on interviews with executives of the innovating firm, executives of a sample of firms using the innovation, and reports and studies made by these firms for internal purposes, it was possible to obtain reasonably reliable estimates of $(P_1 - P_2)$.[5] And once we had an estimate of $(P_1 - P_2)$, it was simple to compute K. Also Q_2 was generally available from published records. Rough estimates of n were obtained from published studies and from the firms. Since K is generally very small, the results are not sensitive to errors in n. Indeed, the expression in equation (8.1) can be approximated quite well in most cases by $(P_1 - P_2)Q_2$, which is the total savings to consumers due to the lower price, if they buy Q_2 units of the product of the industry using the innovation.

To estimate the additional resource saving from the innovation, which equals the bottom shaded area in Figure 8.1 (if the adjustments described in the paragraph before the last are unnecessary), the innovator's profit from the new product was obtained from detailed discussions with the firm's executives, as well as study of relevant financial records. For each year, the costs of marketing and producing the innovation, as well as the costs of carrying out the innovation (R and D, plant and equipment, manufacturing start-up, and marketing start-up),[6] were deducted from the innovator's revenues from the innovation. Note that the R and D costs were adjusted to allow for the fact that the innovator invested R and D resources in uncommercialized R and D projects. To make this adjustment, we obtained estimates from each of the innovating firms of the average number of dollars spent on uncommercialized R and D projects per dollar spent on a commercialized R and D project during the relevant period. Then we multiplied the innovator's R and D outlays (in each year) on the innovation by this number in order to get an estimate of the total R and D investment, including a pro-rated allowance for uncommercialized projects. In cases where the adjustments described in the paragraph before the last were necessary, estimates of the forgone profits from displaced products were deducted, and the profits of imitators were added, to the innovator's profits. These estimates were obtained from the relevant firms.

5. We assume that the cost reduction experienced by the industry using the innovation is passed on to consumers. In view of the market structures of the industries in question, this seems to be a reasonable assumption. Also, the available evidence seems to support this assumption. But it is important to note that if the cost reduction were kept by the industry using the innovation, the social benefit would equal this industry's cost saving due to the innovation plus the innovator's (adjusted) profits. Since the industry's cost saving must be greater than our measure of the upper shaded area in Figure 8.1, our measure of social benefit would be conservative.

6. For discussions of the cost categories involved, see Chapter 4 and E. Mansfield, J. Rapoport, J. Schnee, S. Wagner, and M. Hamburger, *Research and Innovation in the Modern Corporation* (New York: Norton, 1971).

4. Parallel Innovative Efforts, Time Horizon, and Rates of Return

In calculating the social rate of return, we must recognize that besides the innovator, other firms or research organizations in the United States[7] may have invested resources (prior to the introduction of the innovation in question) in R and D and related innovative activities aimed at innovations of essentially the same kind as this one. Clearly, it is not easy to obtain data on the extent of such investments, but fortunately the difficulties seem to be less than they might appear, for two reasons. First, only a limited number of organizations could reasonably have been expected to have been doing R and D in the relevant area, and if they had been devoting any substantial amount of resources to such work in the relevant time frame, it is inconceivable that the current executives of these organizations (and all their competitors) would have been unaware of it. Second, since these innovations occurred some time ago, firms generally are quite willing to discuss whether they were carrying out work of the relevant kind at that time. Moreover, and this is particularly fortunate, our results are quite insensitive to errors in the estimated investment in R and D carried out by others. Even if the true social research expenditures were ten times our estimate, and the true social development expenditures were double our estimate, the results change remarkably little. (To a considerable extent, this is due to the fact, noted in Chapter 4, that R and D costs are often much less than the total cost of launching an innovation.)

Based on interviews with executives of the innovating firm as well as of other firms that could reasonably be expected to do (and be aware of) R and D of the relevant kind, it appears that, in most of the cases in our sample, no other firm or research organization was doing work aimed at roughly the same kind of innovation. Thus, in these cases, the private investment seems to be a good approximation of the social investment. In the remaining cases, other firms or individuals were engaged in R and D aimed at the same kind of innovation. In the cases where this R and D was *unsuccessful,* we obtained as accurate an estimate as possible of the cost of this unsuccessful R and D, and added this figure to the R and

7. Note that we are concerned here only with the costs and benefits arising within the United States. Thus, we ignore the fact that some of these innovations resulted in significant social benefits in other countries. Also, we do not attempt to include whatever costs may have been incurred in other countries in an attempt to produce innovations of roughly the same kind.

Note too that each rate of return estimated here is on the investment in the innovation in question, given that previous investments in basic science, education, and so forth had already occurred. Clearly, the rates of return on the investments considered here were dependent on the previous investments. None of these innovations was a direct outgrowth of developments in government or academic institutions, but work in such institutions may have been helpful in a variety of ways.

D costs described in the previous section to get an estimate of the total social cost of the relevant R and D carried out by the innovator and others.[8]

In the one case (among product innovations used by firms) where other firms or individuals were engaged (prior to the introduction of the innovation in question)[9] in R and D aimed at roughly the same kind of innovation, and where this R and D was *successful,* we must recognize that the innovator's investment only resulted in the innovation's availability at an earlier point in time, not in all of the social benefits from the innovation up to the relevant time horizon. In other words, the proper comparison is between what would have occurred if the innovator had not carried out the innovation (but other firms were free to do so) and what in fact occurred. In this case, we obtained as accurate an appraisal as we could of the date when the innovation would have appeared if the innovator had not carried out the innovation, and we calculated the social benefits only during the period between the date when the innovation occurred and the date when it would have appeared if the innovator had done nothing.[10] Of course, it frequently is very difficult to estimate when the innovation would have occurred if the innovator had not carried it out. But in this particular case a realistic (if perhaps somewhat conservative) estimate is that the second firm to produce the innovation would have come up with it when in fact it did so, regardless of whether the innovator preceded it or not.

For most innovations, these calculations were carried out for each year from the beginning of work on the innovation until 1973. Thus, our estimates of the social benefits are conservative, since all benefits after 1973 are ignored. But in some cases, this would introduce a serious distortion, since the innovation is relatively new. In these cases, forecasts were made of the consumers' surplus and

8. Of course, some of the unsuccessful R and D directed at one innovation may be part of the uncommercialized R and D carried out by the innovator in the case of another innovation. Really, the social cost of an innovation should include a pro-rated share of the cost of uncommercialized R and D carried out by the innovator, excluding the cost of its uncommercialized R and D aimed at innovations where other firms beat it to the punch. But there is no way to estimate the latter cost, and our procedure clearly biases the social rates of return downward.

9. Work of this sort that occurs after the innovation takes place is of a different kind. It is directed at *imitating* or *improving on* the innovation, and is not properly regarded as part of the work *producing* the innovation.

10. Note that this procedure assumes that once the innovator is joined by another firm producing the innovation, they adopt a pricing policy that is just the same as the other firm would have adopted if it were the sole producer of the innovation. Needless to say, this may not be true, since there may be advantages stemming from competition between the two firms. However, we adopt this assumption because it results in a conservative estimate of the social rate of return and because any other assumption would have to be based largely on speculation.

the innovator's profits (adjusted for imitators' profits and for profits on older products) in each year up to 1980. These forecasts are based on firms' expectations concerning $(P_1–P_2)$, Q_2, and the relevant profits in the next few years. They are intentionally very conservative, so whatever bias there is in the resulting rates of return should be downward.

Having made the calculations described in this and the previous section for each year, we have an estimate of the net social benefits (which may be positive or negative) from the innovation for each year. Then we can compute the internal social rate of return, the interest rate that makes the present value of the net social benefits equal to zero. In other words, it is the interest rate, i, that results in the following equality:

$$B(t) + \frac{B(t+1)}{1+i} + \frac{B(t+2)}{(1+i)^2} + \ldots + \frac{B(t+n)}{(1+i)^n} = 0, \tag{8.2}$$

where $B(t)$ is the net social benefit in year t, t is the first year in which the net social benefit is nonzero, and $(t+n)$ is the last year in which the net social benefit is nonzero.

Finally, we also compute the private rate of return from the innovator's investment in each innovation. To do so, we calculate the cash flow to the innovator from the innovation during each year. This calculation involves the subtraction of all costs incurred by the innovator to carry out, produce, and sell the innovation (including the allowance described in the previous section for R and D on uncommercialized projects) from the innovator's revenues from the innovation. Also, profits that the innovator would have earned on products displaced by the innovation must be subtracted.[11] The time period over which these computations were made was generally up to 1973, but in some cases (as in the case of the social rate of return) it extended to 1980. Again, the forecasts in the latter cases are decidedly conservative. The net private benefits in each year, like the Bs in equation (8.2), were deflated. The Consumer Price Index, which generally was used, is not ideal, but it seems very unlikely that the results will be affected in an important way by this choice of a deflator.

5. *Product Innovations Used by Households*

In this section, we turn to the model used to calculate the social benefits for product innovations used by households. Since all of the innovations of this type

11. In all but one case, it is assumed that the profits from the displaced product would have continued (up to the time horizon indicated below) if the innovation had not occurred. This seems reasonable, based on the facts. But in the one remaining case, there is reason to believe that if the innovation had not occurred, another innovation superior to this one would have occurred at a particular point in time afterward. Thus, we assume that the innovator would have earned nothing on the displaced product after this point in time. Of course, it also earned nothing on the innovation after this point.

in our sample are meant to reduce the cost of some particular household activity, basically the same model will apply. Suppose that the demand curve for this activity is DD' in Figure 8.2 and that the innovation reduced the price of performing this activity from P_1 to P_2. Suppose too that the innovator makes a profit of r dollars every time this activity is performed using the product innovation. Then the social benefits from the innovation can be measured by the sum of the two shaded areas in Figure 8.2. These two areas are exactly analogous to those in Figure 8.1, and the methods used to estimate them are similar to those described above, except that no interviews were carried out with users. Instead, the results are based entirely on the estimates and findings of the firms. As in the previous sections, it is necessary to correct the R and D costs for uncommercialized projects, to estimate R and D costs incurred by firms and organizations other than the innovator, to deduct the profit from displaced products from the innovator's profit, and to include the profits of imitators.

However, not everything is the same as in previous sections. For one thing, one of these innovations, the stain remover, resulted in some environmental

FIGURE **8.2**

Social Benefit from Product Innovation that Reduces the Price of a Household Activity

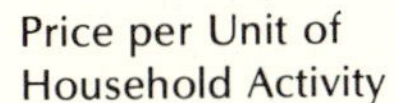

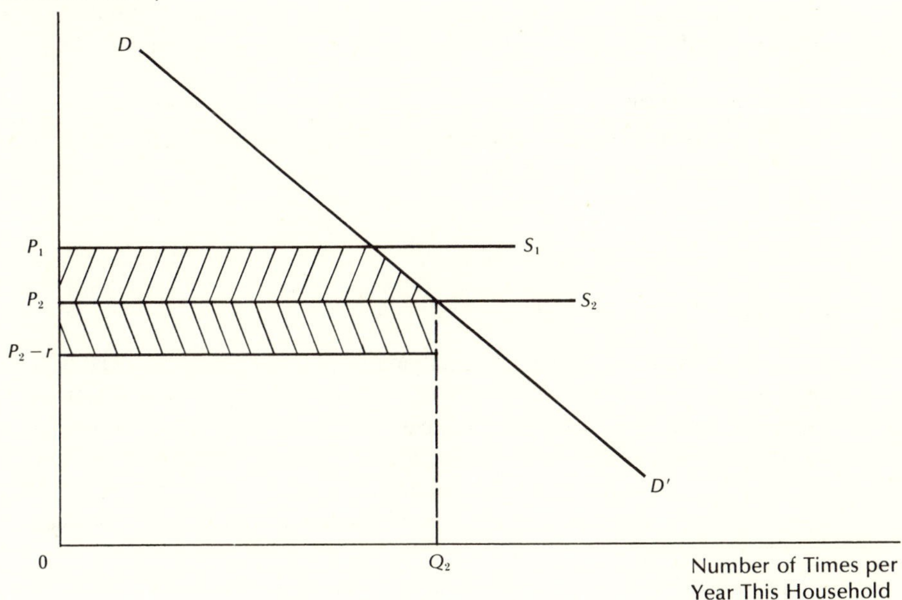

costs, since it increased the amount of a certain chemical that is put into our water supplies. As a rough measure of these costs, we used an estimate (obtained from officials of the Environmental Protection Agency) of the cost of removing a unit of this chemical from water, given that it occurs in a particular concentration. Then we multiplied this cost times the number of units of the chemical that is deposited in the water, given that consumers use the innovation to perform the relevant household activity Q_2 times per year. Finally, this estimate was deducted from the social benefit to correct, at least roughly, for the environmental costs. Note that this was the only innovation in our entire sample that resulted in any substantial environmental costs, so far as we could determine.[12]

In all but one of these cases, there is no reason to believe that if the innovator had not introduced the new product, any other firm would have done so in the relevant time frame. But in the remaining case, a competitor introduced a similar new product only a short while after the innovator introduced its new product. In this case, we obtained estimates from the relevant firms concerning the quantity of its new product that the competitor would have sold each year if the innovator had not developed and introduced its new product. Then we deducted this amount from the actual combined sales of both new products to get the extra amount that was used each year because the new product was introduced earlier than it otherwise would have been, due to the innovator's actions. Then we calculated the consumers' surplus based on this extra amount. Also, we calculated the innovator's profit less the profit that would have been earned by the competitor on the share of the innovator's sales that the competitor would have gotten if the innovator had not introduced its product (and less the profit on products displaced in the remainder of cases). Based on these calculations, we computed the social rate of return.

Finally, as in the case of product innovations used by firms, we computed private as well as social rates of return. The calculations are entirely analogous to those described in the previous section. In all cases, the calculations end with 1973.

6. *Process Innovations*

Next, let's turn to the model underlying our calculation of the social benefits from process innovations. In the case of three of the four process innovations included in our sample, there was no apparent effect on product prices. By lowering the costs of the innovators, these process innovations increased the innovator's profits. Also, since they were imitated (or used at nominal cost) by other firms, they soon increased the profits of other firms as well. The total decrease in

12. Note that some of these innovations had positive environmental effects. However, since these benefits are hard to measure, we ignore them. Of course, whatever bias this introduces results in our estimates of the social rate of return being conservative.

costs (which equals the increase in profits) of all of the relevant firms is a measure of the social benefit of each of these innovations in a particular period. It equals the social saving in resources due to the innovation. For each of these innovations, we estimated the total reduction in costs on the basis of interviews with the relevant firms and of studies and reports concerning the cost reductions due to the innovations and the extent of their utilization.

In the case of the fourth process innovation (the innovation in industrial heating equipment), the situation was different. In this case, the innovator shared the gains with its customers. As shown in Figure 8.3, it reduced the price of its product from P_1 to P_2. Thus, the social benefit equals the sum of the two shaded

FIGURE 8.3

Social Benefit from Process Innovation that Results in Price Reduction by Innovator

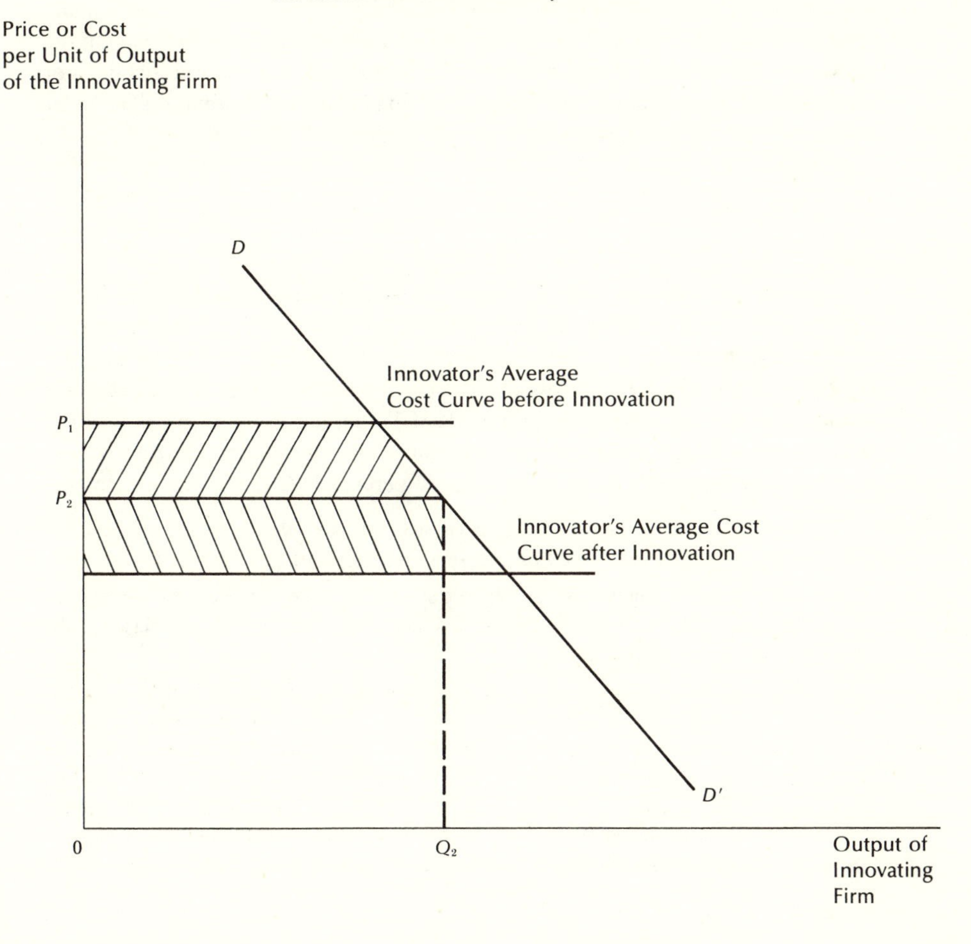

areas, the top shaded area being the consumers' surplus resulting from the price reduction, and the bottom shaded area being the profits to the innovator from the innovation. The situation is almost exactly the same as in Figures 8.1 and 8.2.

For one of the process innovations, it was necessary to estimate the length of time that the innovator hastened the appearance and introduction of the new process. (In the other cases, there is no reason to believe that if the innovator had not introduced the innovation in the relevant time period, any other firm would have done so.) In this case, other firms were doing R and D along somewhat similar lines, and there was some feeling among knowledgeable engineers and managers that the new process might have been developed and introduced about four years later than in fact occurred if the innovator had not done so. In this case, we assume that the diffusion of the new process would have begun four years later (but that it would have proceeded at the same rate) if the innovator had not developed and introduced it. Then the social and private savings are computed by comparing the actual extent of use of the innovation with what it would have been under this assumption. In addition, the social rate of return based on the entire industry's investment in this sort of innovation is estimated, this estimate having the advantage that it is free of any such assumption about how quickly competitors of the innovator would have developed and introduced the innovation.[13] (The latter estimate, which is the lower of the two, is shown in Table 8.1.)

7. Social and Private Rates of Return

The results, shown in Table 8.1, seem to have at least three important implications. First, they indicate that the social rate of return from the investments in these seventeen innovations has been very high. The median estimated social rate of return is about 56 percent. Moreover, for a variety of reasons, these estimates are likely to be conservative lower bounds.[14] To put these results in perspective, recall that Griliches found that the internal social rate of return from hybrid corn—a very successful innovation—was 37 percent.[15] Clearly, the investments

13. For two of the processes, the calculations end with 1973 or 1972; for the other two, forecasts are used up to 1980 (and in one case, beyond).

14. For example, in many cases, we ignore all post-1973 benefits from the innovation. Throughout this study, we have tried to use estimates and procedures that, if biased at all, would be biased in this direction.

15. See Griliches' "Research Costs and Social Returns: Hybrid Corn and Related Innovations," *Journal of Political Economy* (1958). The models used in sections 3 and 4 are in the same spirit as the one applied by Griliches. The principal differences are that we have extended the analysis to include the pricing policies of the innovator, the effect on displaced products, and the costs of uncommercialized R and D and of R and D done outside the innovating organization. Another noteworthy study of this kind is W. Peterson, "Returns to Poultry Research in the United States," *Journal of Farm Economics* (1967),

TABLE **8.1**

Social and Private Rates of Return From Investment in Seventeen Innovations

INNOVATION	RATE OF RETURN (PERCENT)	
	SOCIAL	PRIVATE
Primary metals innovations	17	18
Machine tool innovation	83	35
Component for control system	29	7
Construction material	96	9
Drilling material	54	16
Drafting innovation	92	47
Paper innovation	82	42
Thread innovation	307	27
Door-control innovation	27	37
New electronic device	negative	negative
Chemical product innovation	71	9
Chemical process innovation	32	25
Chemical process innovation	13	4
Major chemical process innovation	56[a]	31
Household cleaning device	209	214
Stain remover	116	4
Dishwashing liquid	45	46
Median	*56*	*25*

[a] Based on investment of entire industry. See section 6.

that have been made in industrial innovations have, on the average, yielded handsome social returns, if these innovations are at all typical.

Second, the results indicate that the private rates of return from the investments in these innovations have been much lower than the social rates of return.[16] The median private rate of return (before taxes) was about 25 percent. In

656–669. This study concludes that the average rate of return from the investment in poultry research was 18 percent.

16. Note that there is nothing in our procedures which makes it inevitable that an innovation's social rate of return exceeds its private rate of return. For a variety of reasons (such as investments in R and D by organizations other than the innovator, and the transfer of profits from the producers of displaced products to the innovator), the social rate of return can be less than the private rate of return—and, in fact, this turns out to be the case for four of the innovations in our sample.

One of the innovations was developed by a nonprofit organization. In this case, the private rate of return is defined as the rate of return to this organization and the licensees that produced and marketed the innovation.

When we compute a confidence interval for the proportion of innovations where the

interpreting this number, it is important to recognize the riskiness of this kind of investment (and the fact that these are pre-tax figures). This riskiness is evidenced by the enormous variation in the private rates of return in Table 8.1. In the case of six innovations, the private rate of return (before taxes) was less than 10 percent, while for five innovations, it was more than 40 percent.

Third, the results indicate that, in about 30 percent of the cases, the private rate of return was so low that no firm, with the advantage of hindsight, would have invested in the innovation, but the social rate of return from the innovation was so high that, from society's point of view, the investment was well worthwhile. We would also like to estimate the extent to which innovations with favorable social rates of return have anticipated rates of return that are so low that they are not carried out. Unfortunately, Table 8.1 cannot provide much information on this score for the obvious reason that all of these innovations were carried out. Also, expected and actual private rates of return may differ considerably. More will be said on this score in Chapter 10.

8. Factors Associated with the Gap between Social and Private Rates of Return

Economists have long been interested in the sorts of innovative activity where the gap between social and private rates of return, if it exists, is particularly large. There has been considerable speculation on this score, but it seems fair to say that the current state of knowledge in this regard is extremely limited. Economic theorists have generally singled out three factors as being of considerable importance in determining the extent of the gap between the social and private rates of return from an innovation. The first of these factors is the market structure of the innovator's industry. If the innovator is faced with a highly competitive environment, it is less likely that it will be able to appropriate a large proportion of the social benefits than if it has a secure monopoly position or if it is part of a tight oligopoly. Of course, the extent to which the innovator is subjected to competition, and how rapidly, may depend on whether the innovation is patented. Another consideration of at least equal importance is how expensive it is for potential competitors to "invent around" the innovator's patents, if they exist, and to obtain the equipment needed to begin producing the new product (or

---

social rate of return exceeds the private rate of return, we find that the probability is 0.90 that this percentage lies between 60 percent and 92 percent. Thus, if we could treat our sample as randomly chosen, we could be quite confident that the social rate of return exceeds the private rate of return in the majority of cases. Also, a t-test indicates that the mean difference between social and private rates of return, which is 47 percent, differs significantly from zero at the .01 level. And so does a sign test, which does not assume normality.

using the new process). In some cases, like du Pont's nylon, it would have been extremely difficult to imitate the innovation (legally). In other cases, a potential competitor could obtain and begin producing a "me too" product (or using a "me too" process) at relatively little cost.

A second factor that economists have emphasized as a determinant of the size of the gap between social and private rates of return is whether the innovation is major or minor. (A reasonable measure of an innovation's importance—at least in our sample—might be the annual net social benefits derived from it.) According to R. C. O. Matthews, the "degree of appropriability is likely to be less . . . in major innovations than in minor ones . . . ,"[17] since major innovations are more likely, in his view, to be imitated quickly. Similarly, on the basis of a model stressing the indivisibility of information, Kenneth Arrow concludes that "the inventor obtains the entire realized social benefit of moderately cost reducing inventions but not of more radical inventions."[18]

A third factor that is sometimes cited is whether the innovation is a new product or process. Thus, Matthews thinks that the degree of appropriability may be less for process innovations than for product innovations. On the other hand, Nelson, Peck, and Kalachek stress that new processes can often be kept secret and that it frequently is difficult for one firm to find out what processes another firm is using.[19] This, of course, suggests that the gap between social and private rates of return might be greater for products than for processes. Neither Matthews nor Nelson *et al.* present any evidence on this score.

Although most of these hypotheses seem quite plausible, the truth is that none of them has been subjected to any sort of systematic empirical tests. Based on the results presented in the previous section, it is possible for the first time to attempt some empirical tests of these hypotheses. These tests are made in two steps. First, we divide the sample into two groups, processes and products, and see whether the average difference between the social and private rates of return is different in one group than the other. Second, looking at products alone (since there are too few processes to support such an analysis), we assume that

$$D_i = \alpha_0 + \alpha_1 S_i + \alpha_2 P_i + \alpha_3 I_i + z_i, \tag{8.3}$$

17. R. C. O. Matthews, "The Contribution of Science and Technology to Economic Development," in B. Williams, ed., *Science and Technology in Economic Growth* (London: Macmillan, 1973), p. 14.

18. K. Arrow, "Economic Welfare and the Allocation of Resources for Invention," in National Bureau of Economic Research, *The Rate and Direction of Inventive Activity* (Princeton: Princeton University Press, 1962), p. 622. Arrow's model bears a resemblance to ours in section 3, but he was interested in a somewhat different set of questions, and no attempt was made to use it as a basis for empirical work.

19. R. Nelson, M. Peck, and E. Kalachek, *Technology, Economic Growth, and Public Policy* (Washington, D.C.: Brookings Institution, 1967), p. 161.

where D_i is the difference between the social and private rates of return from the ith innovation, S_i is the annual net social benefit (in thousands of dollars) from the ith innovation three years after its initial introduction (a crude, but reasonable, measure of the innovation's importance), P_i is a dummy variable that equals 1 if the innovation is patented and zero otherwise, I_i is an estimate of the minimum amount of money (in thousands of dollars) it would have cost any of the innovator's competitors to legally imitate and introduce the ith innovation, and z_i is a random error term. According to the hypotheses stated above, one would expect α^1 to be positive and α^2 and α^3 to be negative. Ordinary regression techniques were used to determine whether this was the case.

The results suggest that the difference between the social and private rates of return tended to be higher for products than for processes. (The average value of D_i was about 57 percent for products and about 21 percent for processes.) But this difference is not statistically significant. Turning to the estimation of the αs, we obtained data concerning I_i for each innovation.[20] Then, combining these data with the information concerning D_i, P_i, and S_i (which stems from the work in previous sections), we used least-squares to estimate the αs, the resulting equation being

$$D_i = \underset{(0.46)}{12.40} + \underset{(4.34)}{0.0313}\, S_i - \underset{(1.87)}{0.0509}\, I_i, \tag{8.4}$$

$$\bar{R}^2 = 0.67$$

where the t statistic of each of the regression coefficients is shown in parentheses, and P_i is excluded because its regression coefficient is not even close to being statistically significant. Despite the relatively small sample size, the estimate of α_1 is highly significant, and the estimate of α_3 is significant at the .05 level. Both have the expected signs.

Thus, the results seem to support the hypotheses that the gap between social and private rates of return tends to be greater for more important innovations and for innovations that can be imitated relatively cheaply by competitors. Apparently, when the cost of imitating the innovation is held constant, it makes little or no difference whether the innovation is patented—which seems reasonable because whether or not a patent exists is of relevance largely (perhaps only) because of its effects on the costs of imitation. It is important to note that S_i and I_i can explain about two thirds of the observed variation in this gap among the

20. These estimates were obtained from the innovating firms. In each case, executives of the innovating firm were asked to estimate the minimum amount that it would have cost a competitor to imitate the innovation (legally). In some cases, the estimate of I_i was as high as \$2 million; in other cases, it was as low as \$3,000. Although these estimates are rough, the executives seemed to think that they were not very wide of the mark.

product innovations in our sample.[21] However, at the same time, it is important to bear in mind the smallness of the sample.

9. Private Returns from a Major Firm's Aggregate Investment in Innovative Activities

There is always the possibility that the results presented in previous sections are biased in some way, since the innovations were not chosen in a strictly random fashion. To check on whether such a bias exists, as well as to extend our previous findings, we obtained data concerning the returns from the innovative activities of one of the nation's largest firms. This firm has made estimates since 1960 of the benefits obtained from its R and D efforts, these estimates being used for internal planning purposes. This firm is among the largest members of an industry that is neither among the most research-intensive nor among the least research-intensive. In terms of the percent of sales devoted to R and D, this firm is reasonably representative of our nation's largest firms.

For each year since 1960, this firm has put together a careful inventory of the technological innovations arising from its R and D and related activities. Then it has made detailed estimates of the effect of each of these innovations on its profit stream. Specifically, in the case of product innovations, the firm computed for each new product the expected difference in cash flows over time between the situation with the new product and without it, including the effect of the new product on its profits from displaced products. In the case of process innovations, it computed the expected difference in cash flow between the situation with the new process and the situation without it, this difference reflecting, of course, the savings associated with the new process. In addition, the firm has updated these estimates each year. In other words, as time has gone on, the firm has revised its estimates of the returns from past innovations. This, of course, is of crucial importance, since it means that the firm's estimates for innovations occurring in the early and middle 1960s are based on a decade or more of actual experience, not just forecasts. The data we use are taken from the latest available revision, which was in 1973.

Besides these data on the private benefits from the firm's technological innovations, figures are also available concerning the firm's expenditures on R and D and related innovative activities each year. Using these cost figures, as well as

21. One innovation had to be excluded from the regression because its private rate of return was incalculable, since the private net benefit in each year was negative. Note too that the ratio of I_i to the innovator's cost in carrying out the ith innovation was tried as an additional explanatory variable in equation (8.4), but it was not close to being significant. Apparently, the absolute, not relative, cost is what matters most. Also, the effect of P_i is not significant, regardless of whether or not I_i is included in the equation.

the figures concerning the total cash flow of benefits stemming from the new products or new processes that came to fruition each year, we can compute the rate of return from the investment that resulted in each year's crop of innovations. It is worthwhile noting that this rate of return is based on the investment in both commercialized and uncommercialized (and successful and unsuccessful) projects.

The results, shown in Table 8.2, are of interest. Never before (to our knowledge) have data of this kind been published. The results indicate at least two things. First, the average rate of return from this firm's total investments in innovative activities during 1960–1972 was about 19 percent, a figure that is not too different from the median private rate of return for the innovations discussed in previous sections. In its internal calculations, this firm regards investments with rates of return exceeding 15 percent as attractive. According to Table 8.2, the average rate of return from the firm's total investment in innovative activities during 1960–1972 exceeded this figure. Second, innovation is a risky activity, and this is reflected in the results. Both for processes and products, the estimated private rate of return fell short of this figure of 15 percent in about 60 percent of the years. The year-to-year variation in the private rate of return seems greater

TABLE **8.2**

Private Rates of Return from Total Investment in Research and Development and Related Innovative Activities, Major Industrial Firm, Process and Product Innovations, 1960–1972

YEAR	BOTH PRODUCTS AND PROCESSES	PRODUCTS	PROCESSES
		(*percentages*)	
1960	31	21	34
1961	9	0	15
1962	7	17	negative*
1963	26	13	30
1964	15	9	18
1965	16	27	−1
1966	25	22	27
1967	11	11	12
1968	2	−1	5
1969	3	13	−15
1970	6	9	3
1971	12	16	10
1972	14	14	14
1960–1972	*19*	*14*	22

* No major process innovations occurred in 1962.

for processes than for products, which may be related to the fact that the average rate of return is higher for processes than for products.

Although these figures are interesting, their limitations should be stressed. For one thing, the firm does not attempt to include in its calculations any innovation where the discounted value of its benefits (the discount rate being 15 percent) is less than $200,000. Since the firm's benefit figures omit the benefits from such minor innovations, the rates of return are almost surely underestimates. Also, the estimates for more recent years are not as reliable as those for the early and middle 1960s. Nonetheless, despite these and other defects in the data, the results seem to provide the most detailed description of a firm's returns from its investments in technological innovation that has been published to date.

10. Social Returns from a Major Firm's Aggregate Investment in Innovative Activities

In the previous section, we were concerned entirely with the private rate of return from the firm's investment in technological innovation. Now let's turn to the social rate of return from this investment. Based on the available data, it is possible only to obtain lower bounds on the social rate of return from the investment leading up to the new products introduced in 1969, and from the investment leading up to all of the processes introduced in 1960–1972.

First, consider the firm's investment that resulted in the new products introduced in 1969. Data are available concerning the external benefits generated by one of the new products introduced in 1969. Specifically, this product allowed firms in another industry to reduce their costs by several millions of dollars per year during the early and middle 1970s. Reasonably accurate data are available from the firm and from the users concerning the extent of these external benefits. Since this product accounted for less than 3 percent of the R and D investment and less than 8 percent of the private benefits obtained from all new products introduced by the firm in 1969, it is a safe bet that it accounted for a fairly small percentage of the external benefits as well. Yet suppose that we make the very conservative assumption that *only* this new product yielded any external benefits at all. In other words, to obtain a lower bound on the social rate of return, let's assume that all of the other new products introduced by the firm in 1969 yielded no external benefits at all. Then, as shown in Table 8.3, the social rate of return would have been about 23 percent, or almost double the private rate of return. Yet, to repeat, this is a very conservative lower bound on the social rate of return.

Second, consider the firm's total investment in the new processes introduced during 1960–1972. Data are available concerning the external benefits generated by three of the new processes introduced during the period. One of these processes, which was a major improvement, was used by other firms in the industry, since its patent protection was too weak to prevent it. The result was a saving of very large amounts by the firm's competitors, as well as substantial savings by the firm itself.[22] The other two processes were of much more modest importance, but they too permitted other firms in the industry to reduce their costs. Suppose that we make the very conservative assumption that *only* these three new processes yielded any external benefits at all. In other words, to obtain a lower bound on the social rate of return, let's assume that all of the other new processes introduced during 1960–1972 yielded no external benefits at all. Then, as shown in Table 8.3, the social rate of return would have been about 55 percent, or well over double the private rate of return. And, to repeat, this estimate of the social rate of return is a very conservative lower bound.

TABLE **8.3**

Private Rate of Return and Lower Bound on Social Rate of Return, Major Industrial Firm, New Products in 1969 and New Processes in 1960–1972

NATURE OF INVESTMENT BY FIRM	PRIVATE RATE OF RETURN	LOWER BOUND ON SOCIAL RATE OF RETURN
New products in 1969	13	23
New processes in 1960–1972	22	55

11. Unemployment, Repercussions on Other Markets, and Future Changes in Technology

Before concluding this chapter, we must take up several additional factors that have not been mentioned explicitly in previous sections. First, new technology can, of course, result in prolonged and serious unemployment of particular kinds of labor. Recognizing this fact, we tried to determine whether there was any substantial unemployment (or change in job content) resulting from each innovation described in section 2. In no case could we find any evidence of an appreciable change of this sort: although labor requirements sometimes were reduced by the innovation, growth in output offset these reductions. Thus, in this sample at least, this factor does not seem to be of substantial importance. Obviously, however, it can be of importance in other cases; and in situations where it is of importance, one must adjust the social benefits, as calculated in previous sections, for the fact that (for a certain period of time) some of the resources saved by the innovation do not find employment elsewhere.[23]

22. This is a good illustration of how difficult it frequently is for an innovator to prevent other firms from imitating its new technology. More will be said on this score in several parts of Chapter 9.

23. See A. Schmitz and D. Seckler, "Mechanized Agriculture and Social Welfare: The Case of the Tomato Harvester," *American Journal of Agricultural Economics* (November 1970).

Second, new technology can, of course, have an effect on other industries and markets besides the ones directly involved. For example, by reducing the cost and price of the product using an innovation, the innovation affects the outputs of goods that are substitutes or complements of this product. However, so long as the prices of these and other goods (other than the product using the innovation) are not materially affected, the effects of these output changes are already included in the measures specified above.[24] In our sample, there is no evidence that any of these innovations had any appreciable impact on any price other than that of the good using the innovation (in section 4) or of the household activity in question (in section 5) or of the good based on the one relevant process (in section 6). Also, one must assume that inputs that were displaced from industries producing goods competing with the product using the innovation readily found employment elsewhere. There is no evidence that this was not the case.

Third, new technology can, of course, play an important role in suggesting and prompting further innovations. Without question, practically all of the innovations in our sample (and in the firm discussed in sections 9 and 10) had this effect. For example, the new type of metal discussed in section 2 undoubtedly will suggest further innovations in the same and related fields. Thus, since each of these innovations is instrumental in bringing about future innovations, each results in social benefits above and beyond what we have measured. However, each of these innovations may have been suggested, prompted, or aided by previous innovations, and there is no feasible way to estimate the extent to which there is a net understatement of benefits.

12. Summary and Conclusions

In this chapter, we have made estimates of rates of return for 17 innovations in a variety of industries, as well as for the R and D portfolio of one of the nation's largest firms. These seem to be the first such estimates that have been carried out for industrial innovations. The results indicate several things. First, they indicate that the social rates of return from these investments in new technology have been very high. For the 17 innovations, the median estimated social rate of return is about 56 percent. For the large industrial firm, a lower bound on the social rate of return from its investment in new process technology in 1960–1972 was 55 percent. Clearly, the investments that have been made in industrial innovations have, on the average, yielded handsome social returns, if these innovations are at all typical.

Second, the results indicate that the private rates of return from the investments in these innovations have been much lower than the social rates of return. For the 17 innovations, the median estimated private rate of return was about 25

24. See E. Mishan, *Economics for Social Decisions* (New York: Praeger, 1972), chap. 6.

percent. For the large industrial firm, the estimated private rate of return from its investment in new process and product technology in 1960–1972 was about 19 percent.[25] In interpreting these figures, one must bear in mind the riskiness of this kind of investment and the fact that they are pre-tax rates of return. This riskiness is evidenced by the enormous variation among the private rates of return from the 17 innovations. For 6 of these innovations, the private rate of return (before taxes) was less that 10 percent, while for 5 of them, it was more than 40 percent.

Third, the results indicate that, for about 30 percent of the innovations in our sample, the private rate of return was so low that no firm, with the advantage of hindsight, would have invested in the innovation, but the social rate of return from the innovation was so high that, from society's point of view, the investment was well worthwhile. Also, it appears that the difference between the social and private rate of return was related to the economic importance of the innovation (measured by absolute annual benefits) and to the costs of imitating the innovation.

We believe that these results go far beyond anything attempted heretofore in this area. Yet their limitations should be borne in mind. Strictly speaking, our sample of innovations is not a random sample, the data sometimes are rough, and for a few of the innovations they are based partly on forecasts. Further, the models that we use are simplified in many respects.[26] For these and other reasons, our results should be treated with caution. The measurement of social and private rates of return from investments in new technology is an extremely difficult business, which is one good reason why so few such measurements have been attempted. Nonetheless, if sensible decisions are to be made regarding public policy toward civilian technology, it is essential that we make the best estimates we can.

25. Of course, for reasons pointed out above, this is an underestimate, but it seems doubtful that the private rate of return exceeds this figure by a great deal.

26. An important point to bear in mind is that the models used here cannot cope with innovations that result in entirely new product characteristics unless they can be expressed in terms of some form of cost reduction. Based on this sample, it appears that models of this sort are widely applicable, but, particularly for very major new consumer products, it is unlikely that they are serviceable, at least in their present form.

9 THE CASE STUDIES: A CLOSER LOOK AT THE BENEFITS AND COSTS OF INDUSTRIAL INNOVATIONS*

1. Introduction

In the previous chapter, we estimated social and private rates of return for seventeen industrial innovations. Although the previous chapter described the basic models on which these results were based, it did not discuss the individual case studies in any detail. Because these case studies are the first of their kind[1] and because the application of the methodologies presented in the previous chapter will be clarified by some illustrative examples, it seems worthwhile to present brief descriptions of some of the case studies. This material should provide insight into the innovation process as well as into the methods used in the previous chapter.

In this chapter, we present brief summaries of seven of the case studies. The data underlying twelve of the case studies are quite reliable, so far as we can tell, while the data underlying the remaining five are distinctly weaker, although still reasonably useful. The seven case studies summarized in this chapter were cho-

* George Beardsley is a co-author of this chapter. He was responsible in particular for sections 4 and 8.

1. As noted in the previous chapter, the only previous case studies of this sort were a few pertaining to agriculture. For example, Griliches and Peterson studied hybrid corn and poultry R and D. Nothing of this sort has been done before for industry.

sen more or less at random from among those where the data are relatively strong. These seven innovations occurred in the metal, thread, chemicals, household cleaner, and industrial equipment industries. Three are industrial products, two are consumer products, and two are processes.

2. *Industrial Product A*

This is a new type of metal. The main users of the new product are manufacturers of home appliances, which constitute about 80 percent of the market for the new type of metal. This new type of metal made it possible to make these appliances more cheaply. This innovation was based on a chance discovery in the laboratories of the innovating firm. The new product was introduced on a developmental basis in the late 1950s and on a commercial basis in 1960. The research and development was done mainly in 1954–1956, 1957–1959 being a period of market development. Since 1960, there has been continuing R and D on modifications. According to the innovator, the costs were as follows: applied research, $45,000; preparation of product specifications, $40,000; pilot plant, $50,000; manufacturing facilities and start-up, $4,000,000; and marketing start-up, $250,000. An additional annual cost of $60,000 from 1960 to 1974 went for R and D on modifications.

The gross profits earned by the innovator are shown for each year in Table 9.1. All marketing (and production) costs pertaining to the new product have, of course, been deducted out. These figures are only approximate, but they should be good enough for present purposes. Note that they take into account the effect of the innovation on the sales of the innovator's other products. Table 9.1 also corrects the innovator's costs for the investment in uncommercialized R and D. Here and below the innovator provided data concerning the ratio (during the relevant period) of total expenditure on R and D to total expenditure on commercialized R and D. Then the R and D costs of the innovation were multiplied by this ratio. Finally, the last column of Table 9.1 shows the net profit to the firm each year from the innovation.

Turning to the social returns from the innovation, the first thing to note is that because the innovator's patent position was weak and the technology was relatively easy to imitate, the innovator's competitors quickly came out with their own versions of the innovation. Based on discussions with these firms (and the innovator), it is possible to estimate reasonably accurately the total increase in profit of the entire industry (of which the innovator is a member) over what it would have earned if the innovation had not occurred. This estimate is shown for each year in the first column of Table 9.2. Note that this profit is gross of the innovator's costs of developing and implementing the innovation, but net of the imitators' costs involved in imitating it. Note too that the industry's total extra profit from the innovation is much less than the innovator's net profit from it (in Table 9.1) because the innovation enabled the innovator to gain considerable

TABLE **9.1**

Net Private Returns from Industrial Product A

YEAR	INNOVATOR'S GROSS PROFIT, INCLUDING EFFECT ON ITS SALES OF OTHER PRODUCTS	COSTS OF INNOVATION TO INNOVATOR	COSTS, CORRECTED FOR UNCOMMERCIALIZED R AND D[a]	INNOVATOR'S NET PROFIT
1954	0	$1,378,000	$1,783,000	-$1,783,000
1955	0	1,378,000	1,783,000	-1,783,000
1956	0	1,378,000	1,783,000	-1,783,000
1957	0	83,000	83,000	-83,000
1958	0	83,000	83,000	-83,000
1959	0	83,000	83,000	-83,000
1960	$1,500,000	60,000	60,000	1,440,000
1961	1,654,000	60,000	60,000	1,594,000
1962	1,808,000	60,000	60,000	1,748,000
1963	1,961,000	60,000	60,000	1,901,000
1964	2,115,000	60,000	60,000	2,055,000
1965	2,269,000	60,000	60,000	2,209,000
1966	2,423,000	60,000	60,000	2,363,000
1967	2,576,000	60,000	60,000	2,516,000
1968	2,730,000	60,000	60,000	2,670,000
1969	7,884,000	60,000	60,000	7,824,000
1970	8,038,000	60,000	60,000	7,978,000
1971	8,192,000	60,000	60,000	8,132,000
1972	8,345,000	60,000	60,000	8,285,000
1973	8,500,000	60,000	60,000	8,440,000

[a] Because of the nature of the R and D in many of the later years, the R and D costs in these years were not corrected for uncommercialized R and D. Obviously, this makes practically no quantitative difference, but it seemed more nearly correct.

business at the expense of its competitors. Further, it should be noted that the profits of an engineering company that supplied some equipment used to help make the new product must also be included in the first column of Table 9.2.

Based on discussions with users and makers of the new product, we estimated the size of the cost reduction in each year for each appliance using the new metal. These figures seem reasonably firm. Then we obtained data concerning the percentage of appliances using the new product in 1973; and assuming linear growth in this percentage (from zero in 1960), we could calculate the value of $(P_1 - P_2)Q_2$ each year on the basis of published output figures for the relevant appliances.[2] The results are shown in the second column of Table 9.2. These results seem quite reliable, based on everything we could find out.

2. For definitions of P_1, P_2, Q_2, and other symbols used in sections 2 through 4 of this chapter, see sections 3 and 4 of chapter 8.

TABLE 9.2

Social Returns from Industrial Product A

YEAR	INDUSTRY'S PROFITS MINUS PROFITS FROM DISPLACED PRODUCT[1]	$(P_1 - P_2) Q_2$	CONSUMERS' SURPLUS	NET SOCIAL BENEFIT
1954	—	—	—	−$1,783,000
1955	—	—	—	−1,783,000
1956	—	—	—	−1,783,000
1957	—	—	—	−83,000
1958	—	—	—	−83,000
1959	—	—	—	−83,000
1960	$1,167,000	—[2]	—	1,107,000
1961	1,667,000	$170,000	$168,000	1,775,000
1962	1,667,000	374,000	370,000	1,977,000
1963	1,667,000	608,000	602,000	2,209,000
1964	1,667,000	870,000	861,000	2,468,000
1965	1,967,000	1,157,000	1,145,000	3,052,000
1966	1,100,000	1,397,000	1,383,000	2,423,000
1967	1,375,000	1,584,000	1,568,000	2,883,000
1968	1,625,000	2,044,000	2,024,000	3,589,000
1969	1,875,000	2,175,000	2,153,000	3,968,000
1970	2,150,000	2,193,000	2,171,000	4,261,000
1971	2,400,000	2,509,000	2,484,000	4,824,000
1972	2,750,000	2,941,000	2,912,000	5,602,000
1973	3,000,000	3,204,000	3,172,000	6,112,000

[1] Also, profits of an engineering company that installed some equipment connected with the innovation are included.

[2] The amount, which was small, is assumed for simplicity to be zero.

If the cost savings are passed along to consumers, the resulting consumers' surplus equals $(P_1 - P_2)Q_2 \times (1 - Kn/2)$. Since the cost reduction per appliance is about $0.20, the price of the relevant appliance is about $200.00, and the price elasticity of demand is about 2 (according to Harberger), the resulting consumers' surplus is about 99.9 percent of the value of $(P_1 - P_2)Q_2$. To be conservative, we multiply $.99 \times (P_1 - P_2)Q_2$, the result being shown in the third column of Table 9.2. And since there is no evidence that any firm or research organization other than the innovator was carrying out work aimed at a similar kind of innovation, the innovator's investment can be taken as an adequate measure of the social costs of producing the innovation. Thus, the last column of Table 9.2, which is column 1 of Table 9.2 plus column 3 of Table 9.2 minus column 3 of Table 9.1, shows the net social benefit each year from the innovation.

Note that these social-benefit figures are conservative on several counts. First, they ignore whatever savings occurred in industries other than appliances due to the innovation. (Recall that appliances account for only 80 percent, not all, of the market for the new product.) Second, they ignore benefits outside the United States. Third, they ignore benefits after 1973. Fourth, they ignore the fact that the innovation, because it was manufactured differently than the product it displaced, had some beneficial effects on working conditions for labor.

Based on the figures in Table 9.2, the nominal internal rates of return for this innovation were 20 percent (social) and 21 percent (private), the corresponding real rates of return being 17 percent and 18 percent.

3. *Industrial Product B*

This is a new type of thread. One important advantage of this thread is that it allows higher sewing-machine speeds, thus reducing the costs of garment manufacturers. The R and D leading up to this innovation was carried out by the innovator in 1962–1964, and the innovation was put on the market in 1965. Since that time, the innovator has continued to invest in R and D concerning the innovation. Table 9.3 shows the amount spent each year by the innovator on R and D and related activities concerning the innovation.

The profits (gross of R and D costs) derived from the innovation by the innovator are also shown in Table 9.3. These profits are net of whatever profits the innovator would have made on its products that were displaced by the innova-

TABLE **9.3**

Private Benefits from Industrial Product B

YEAR	GROSS PROFITS	R AND D AND RELATED COSTS	DEFLATED NET PROFITS (ADJUSTED FOR UNCOMMERCIALIZED R AND D)
1962	—	$25,000	−$50,000
1963	—	37,000	−73,000
1964	—	49,000	−96,000
1965	$135,000	74,000	3,000
1966	175,000	75,000	26,000
1967	216,000	85,000	46,000
1968	259,000	90,000	76,000
1969	328,000	100,000	117,000
1970	379,000	100,000	154,000
1971	432,000	75,000	232,000
1972	488,000	50,000	310,000
1973	546,000	50,000	260,000

tion. The profit figures were computed from sales data and profit rates provided by the innovator. To obtain the net private benefits, we deduct column 2 (as well as a prorated share of uncommercialized R and D) from column 1. The result (expressed in 1967 dollars) is shown in the last column of Table 9.3. The prorated share of uncommercialized R and D is figured on the basis of the firm's estimate of the number of dollars it spent on uncommercialized R and D per dollar of commercialized R and D during the relevant time frame.

To estimate the social benefits from the innovation, we must estimate the savings to garment producers due to the innovation. Based on data concerning the percent of garment-manufacturing costs that are machine-operating costs and the percentage reduction in machine-operating costs due to the innovation, it is possible to compute the percentage reduction, on the average, of garment-production costs due to the innovation (net of the higher cost of the new thread). Based on the sales of the new product, it is possible to estimate the total amount of garment production using the innovation. Thus, assuming that the cost reduction is passed on to the consumer, one can calculate $(P_1 - P_2)Q_2$, which is shown in Table 9.4. To obtain consumers' surplus, one must multiply $(P_1 - P_2)Q_2$ by $(1 - \frac{1}{2}Kn)$. The value of K varies from year to year due to changes in thread prices, its average value being about 0.0165. As for n, Houthakker and Taylor do not find a price elasticity of demand for clothing significantly different from zero, but to be conservative, we assume n to be 1. Thus, we multiplied $(P_1 - P_2)Q_2$ by $(1 - K/2)$, the result being shown in the second column of Table 9.4.

To calculate the social returns from the innovation, we must also recognize that the innovation undoubtedly reduced the profits of competitors, since the in-

TABLE **9.4**

Social Benefits from Industrial Product B

YEAR	$(P_1 - P_2)Q_2$	CONSUMERS' SURPLUS	FORGONE PROFITS BY COMPETITORS	DEFLATED NET SOCIAL BENEFITS
1962	—	—	—	−$50,000
1963	—	—	—	−73,000
1964	—	—	—	−96,000
1965	$3,363,000	$3,336,000	$66,000	3,443,000
1966	4,354,000	4,319,000	170,000	4,295,000
1967	5,411,000	5,368,000	209,000	5,205,000
1968	6,521,000	6,469,000	251,000	6,043,000
1969	7,696,000	7,634,000	328,000	6,770,000
1970	8,930,000	8,859,000	379,000	7,445,000
1971	10,227,000	10,145,000	432,000	8,240,000
1972	11,582,000	11,489,000	488,000	9,090,000
1973	13,000,000	12,896,000	546,000	9,614,000

novation displaced older types of thread. According to estimates made by the innovator, the profits that competitors would have earned on the products that were displaced by the innovation are as indicated in the third column of Table 9.4. Note that most of the innovator's profits were at the expense of its competitors. Apparently, for most years, the profits earned on the innovation were equal to the amount that would have been earned on displaced products. Since there is no evidence that any other firm or research organization was carrying out R and D aimed at a similar kind of innovation, the R and D costs in Table 9.3 seem to be a good estimate of the social R and D costs. However, it was easy and cheap for competitors to imitate the innovation; and within six months after the innovation's introduction, competitors began imitating it.

To calculate the net social benefits each year, we must subtract column 3 of Table 9.4 from column 2, deflate the difference, and add it to the last column of Table 9.3. The result is shown in the last column of Table 9.4. Based on the figures in Tables 9.3 and 9.4, the real internal rates of return were 307 percent (social) and 27 percent (private). These estimates are conservative because they omit returns achieved after 1973.

4. Industrial Product C

This is an intermediate good which is sold in bulk. It is a product satisfying an industrial need that has been met by the innovating firm ever since it was founded. However, product C has special properties which make it quite superior in performance compared to its predecessors under certain conditions of use. The usefulness of product C is derived from its application in conjunction with heavy capital equipment units in the transportation industry.

The origins of product C lie in the evolutionary-type R and D programs that the innovating firm carries on in areas of continuing interest to itself and to the industry. The R and D program specifically associated with product C began in 1967. The technical breakthrough which allowed product C to become commercially feasible—i.e., profitably marketable—occurred in 1969. In 1970 and 1971 additional technical accomplishments were recorded which reduced the cost of production of product C. A further accomplishment, a relatively minor modification of sorts, was recorded in 1972. The innovator's profit—gross of the investment in the innovation—has been estimated in detail by the firm as of 1973. These values, on a year-by-year basis, are shown in the next-to-last column in Table 9.5. These figures allow for any loss of sales to substitute products which the innovating firm offered prior to the introduction of product C.

Having estimated the gross value to the firm, it is also necessary to identify the associated costs. The innovating firm was able to provide an estimate of the applied R and D program costs, which run from 1967 through 1972, and include prorated charges for the overhead groups of planning and plant general services.

TABLE 9.5

Estimated Income Stream to the Firm Resulting from Its Investment in Industrial Product C

YEAR	ADJUSTED R AND D COSTS	IMPLEMENTATION COSTS	GROSS PROFITS	NET PROFITS [a]
		(*thousands of dollars*)		
1960	$1	—	—	−$1
1961	1	—	—	−1
1962	3	—	—	−3
1963	6	—	—	−6
1964	8	—	—	−8
1965	10	—	—	−10
1966	10	—	—	−10
1967	220	$120	—	−340
1968	200	110	—	−300
1969	210	120	—	−330
1970	270	150	$50	−380
1971	140	80	330	110
1972	140	80	550	330
1973	—	—	630	630
1974	—	—	650	650
1975	—	—	640	640
1976	—	—	530	530
1977	—	—	40	40
1978	—	—	60	60
1979	—	—	20	20

[a] Because of rounding, net profits may not always equal gross profits less costs.

In addition, however, there was surely a cost to the firm from the more fundamental research associated with the development of product C. (Since research of this sort could not be identified as pertaining to a specific product, it could not be recorded as such.) To deal with this problem, we assume that the maximum percentage of the whole annual applied product R and D group budget that was accounted for by the particular project in question in any year was the percentage of all of the more fundamental, or "basic,"[3] research expense which is attributable to that particular project for each and every year that relevant "basic" research expenditures were carried out. For product C this figure is only 3 percent. This assumption is almost certainly conservative, in that it overestimates these

3. Here and below, "basic" is used to characterize research that the firm designates as basic, although, by the National Science Foundation definition, it may not be the proper designation.

costs. The resulting estimate of the total R and D costs each year attributable to the innovation is shown in the first column in Table 9.5.

There are other costs yet to be considered. These are the implementation costs which go for the multiplicity of corporate efforts outside the R and D groups that are necessary in order to utilize a new technology and to realize any subsequent commercial gains from it. These efforts occur in marketing, patent administration, engineering, manufacturing, and other varied groups, and are most difficult to pinpoint. Based on intensive discussions with the firm, the implementation costs associated with product C are estimated to be two thirds of all applied R and D costs associated with product C less the total estimated "basic" research costs. The firm views this as an upper bound for the implementation costs. These costs are distributed over the applicable years of applied R and D costs, using those costs to weight the distribution.[4]

Another adjustment is made in attempting to examine the income stream identified with the firm's investment in product C. According to an earlier survey, 72 percent of the innovator's applied R and D dollars are estimated to be spent on programs successfully completed and available for commercial exploitation. Thus, all applied R and D costs, as well as implementation costs (which are estimated on the basis of applied R and D costs), have been multiplied by 1/(0.72) to reflect a fair share of the cost of those programs which do not reach commercialization. Table 9.5 shows, year by year, the components of the net-profit stream to the innovating firm resulting from its investment in the development and commercial introduction of product C.

Next, we must consider the reduction in the costs of the industry using the innovation. The market for product C can easily be divided into two groups of users, each being distinctly separate segments of the transportation industry. Each uses the same particular capital equipment in conjunction with product C, but under very different conditions. The savings associated with the introduction of product C are known fairly well for one industry segment, but not at all for the other.

Several years ago, a field study was conducted by the innovating firm in conjunction with an important user of product C. A formal report was prepared which quantified the saving to the user due to product C. This report took into account the best alternative product and allowed for any price differential. The estimated saving to this single user, as ascertained in these field tests, was based on the number of units of capital equipment which are complemented by product C (or its alternative product) and the circumstances of its usage. These environmental circumstances are important in their relationship to the differential value of product C over its best market alternative. They, in fact, limit the feasible market for this new product.

4. In most other case studies, inplementation costs are not broken out, and treated separately. However, it is important to note that these costs are included in the analysis in each and every case.

By 1973, the company which had participated in the field test, a second major customer, and several smaller customers were users of product C. It is anticipated that they will remain the only users within this distinct segment of the user market. The innovating firm has managed to estimate for purposes of this study the annual saving to all these users during the anticipated life of product C; the results are shown in Table 9.6. The innovator assumed that product C would be displaced by 1977, so the figures extend only to 1976. These estimates are considered reliable.

TABLE **9.6**

Estimated Saving to Subgroup of Users of Industrial Product C

YEAR	SAVING TO CUSTOMER NO 1	SAVING TO CUSTOMER NO. 2	SAVING TO ALL SMALL CUSTOMERS	TOTAL[a]
		(millions of dollars)		
1970	$1.5	—	—	$1.5
1971	1.5	$1.2	—	2.7
1972	1.5	1.2	$0.4	3.1
1973	1.5	1.2	0.3	3.1
1974	1.5	1.2	0.2	3.0
1975	1.5	1.2	0.2	2.9
1976	1.5	1.2	0.1	2.8

[a] Because of rounding, individual items may not sum to totals.

Given the market structure of the industry using the innovation, it is extremely likely that these savings will be passed along to consumers. But to get from the savings in each year to the consumers' surplus in that year, one must multiply these savings by $(1 - Kn/2)$. However, based on published estimates of the price elasticity of demand and the value of K, this factor is so close to 1 that no adjustment is worthwhile. Also, there was no evidence that other firms or research institutions were trying to develop an innovation like product C, so no adjustment seems necessary on that count. Further, according to the innovator, there were no profits from displaced products to be considered.

We calculate the social rate of return from the investment leading to product C, based on two alternative assumptions. First, we assume that the total consumers' surplus arose through the cost saving to the class of users in Table 9.6. Second, we assume that this consumers' surplus is half of the total, since the known users represent only about one half of product C's customers. The relative size of this known share of the market is 52 percent, 42 percent, and 49 percent from 1970 to 1972 respectively, with a weighted average of 47 percent based on volume purchased. Even though the product is used in conjunction with nearly identical equipment in the transportation-market sectors, the different conditions of usage restrict what can be said about the sector where user benefits are an unknown quantity. However, if one considers twice the user benefits to the known subset of customers as the magnitude of total user benefits, then one will probably be closer to the truth than if the second sector is ignored. Table 9.7 shows the stream of annual net social benefits based on each assumption.

Based on these data, the nominal internal rates of return are 78 percent (social, based on first assumption), 102 percent (social, based on second assumption), and 14 percent (private). The corresponding real rates of return are 71 percent, 94 percent, and 9 percent. Thus, the real private rate of return was about 9 percent, and the real social rate of return was at least 71 percent, probably considerably more. In Table 8.1 in the previous chapter, we used the lower of the two social rates of return, which, of course, is in accord with our practice of being as conservative as possible in this regard.

TABLE **9.7**

Annual Net Social Benefits Associated with the Investment Leading to Industrial Product C

YEAR	INNOVATOR'S NET PROFITS	CONSUMERS' SURPLUS BASED ON KNOWN VALUE TO USER SUBGROUP	NET SOCIAL BENEFITS BASED ON FIRST ASSUMPTION	NET SOCIAL BENEFITS BASED ON SECOND ASSUMPTION
		(thousands of dollars)		
1960	−$1	—	−$1	−$1
1961	−1	—	−1	−1
1962	−3	—	−3	−3
1963	−6	—	−6	−6
1964	−8	—	−8	−8
1965	−10	—	−10	−10
1966	−10	—	−10	−10
1967	−340	—	−340	−340
1968	−300	—	−300	−300
1969	−330	—	−330	−330
1970	−380	$1,500	1,160	2,690
1971	110	2,700	2,850	5,580
1972	330	3,100	3,460	6,590
1973	630	3,100	3,690	6,740
1974	650	3,000	3,630	6,600
1975	640	2,900	3,530	6,400
1976	550	2,800	3,350	6,160
1977	40	—	40	40
1978	60	—	60	60
1979	20	—	20	20

Note: Because figures are rounded, individual items may not sum to totals.

5. Consumer Product A

This is a new type of household cleansing device. It reduces the cost of a particular household activity, based on studies that have been made by independent testing agencies. The innovating firm carried out the R and D leading to this product in 1959 and 1960, and introduced the product commercially in 1961. As can be seen in Table 9.8, the innovator has continued to invest in R and D concerning the innovation in the period since its introduction. Column 2 of Table 9.8 shows the amount spent each year by the innovator on R and D and related costs pertaining to this innovation. Column 1 of Table 9.8 shows the innovator's net profit (net of its R and D and related costs). When this net profit is adjusted for the cost of uncommercialized R and D and deflated, the innovator's deflated profit (in 1958 dollars) is shown in the last column of Table 9.8.

To estimate the social benefits from the innovation, we must know how much the innovation saved the consumer each time it was used, relative to the best alternative. Based on available data, the total saving to consumers—$(P_1 - P_2)Q_2$—was as shown in column 1 of Table 9.9. This figure was derived by estimating the reduction in the cost of each unit of the relevant household activity, and by multiplying it times the postinnovation amount of such activity. Based on these

TABLE **9.8**

Private Returns from Consumer Product A

YEAR	INNOVATOR'S PROFIT (AFTER R AND D AND RELATED COSTS)	INNOVATOR'S R AND D AND RELATED COSTS	INNOVATOR'S DEFLATED NET PROFIT (AFTER ADJUSTMENT FOR UNCOMMERCIALIZED R AND D)
1959	−$34,000	$34,000	−$93,000
1960	−40,000	40,000	−108,000
1961	76,000	44,000	−2,000
1962	1,859,000	101,000	1,587,000
1963	5,536,000	81,000	5,030,000
1964	6,407,000	101,000	5,724,000
1965	4,149,000	114,000	3,558,000
1966	5,285,000	76,000	4,521,000
1967	6,373,000	24,000	5,383,000
1968	6,758,000	20,000	5,512,000
1969	5,623,000	82,000	4,331,000
1970	5,873,000	73,000	4,295,000
1971	5,250,000	43,000	3,682,000
1972	4,268,000	114,000	2,855,000
1973	4,492,000	115,000	2,855,000

TABLE **9.9**

Social Returns from Consumer Product A

YEAR	$(P_1 - P_2)Q_2$	DEFLATED CONSUMERS' SURPLUS	INNOVATOR'S DEFLATED NET PROFIT[1] LESS FORGONE DEFLATED PROFITS BY COMPETITORS ON DISPLACED PRODUCTS	DEFLATED NET SOCIAL BENEFITS
1959	—	—	−$93,000	−$93,000
1960	—	—	−108,000	−108,000
1961	$215,000	$172,000	−2,000	170,000
1962	1,806,000	1,422,000	397,000	1,819,000
1963	2,267,000	1,762,000	1,258,000	3,020,000
1964	2,354,000	1,803,000	1,431,000	3,234,000
1965	1,896,000	1,424,000	890,000	2,314,000
1966	1,827,000	1,336,000	1,130,000	2,466,000
1967	1,776,000	1,258,000	1,346,000	2,604,000
1968	1,776,000	1,210,000	1,378,000	2,588,000
1969	1,606,000	1,044,000	1,083,000	2,127,000
1970	1,586,000	976,000	1,074,000	2,050,000
1971	1,632,000	960,000	921,000	1,881,000
1972	1,540,000	878,000	714,000	1,592,000
1973	1,547,000	837,000	714,000	1,551,000

[1] Adjusted for uncommercialized R and D.

data, $K = 1/3$, and a conservative estimate of n is 1.[5] Thus, we multiply the value of $(P_1 - P_2)Q_2$ by $(1 - 1/2 \times 1/3) = 0.83$ to get the consumers' surplus. The deflated value of this consumers' surplus is shown in column 2 of Table 9.9.

We must also take account of the fact that the innovation resulted in lower profit for competitors whose products were displaced. If we deduct the deflated profit reduction to competitors (estimated in detail by the innovating firm) from the last column of Table 9.8 we obtain the result in column 3 of Table 9.9. Since there is no evidence that any other firm or research organization was doing R and D aimed at a similar sort of innovation, the cost data in Table 9.8 seem to be a proper measure of social R and D costs. Thus, the deflated net social benefits from the innovation equal the sum of column 2 and column 3 of Table 9.9, which is shown in the last column of Table 9.9.

Based on the figures in Tables 9.8 and 9.9, the real internal rates of return are 209 percent (social) and 214 percent (private). Since benefits arising after 1973 are omitted, these rates of return are somewhat conservative.

5. For definitions of P_1, P_2, Q_2, K, n, and other symbols used in sections 5 and 6 of this chapter, see section 5 of chapter 8.

6. Consumer Product B

This is a new type of stain remover. It reduced the cost to the consumer of removing spots. The innovating firm carried out the R and D leading to the innovation in 1967, and introduced the innovation commercially in 1968. As indicated in Table 9.10, the firm continued to spend money on R and D related to the innovation after its introduction. The innovator's profits (net of its R and D and related costs) from the innovation are shown in the first column of Table 9.10. After adjusting these profits for the cost of uncommercialized R and D, and deflating them, the resulting net profits (in 1958 dollars) are shown in the last column of Table 9.10.

TABLE **9.10**

Private Returns from Consumer Product B

YEAR	INNOVATOR'S PROFIT (AFTER R AND D AND RELATED COSTS)	INNOVATOR'S R AND D AND RELATED COSTS	INNOVATOR'S DEFLATED NET PROFIT (AFTER ADJUSTMENT FOR UNCOMMERCIALIZED R AND D)
1967	−$164,000	$164,000	−$387,000
1968	−10,916,000	0	−8,926,000
1969	533,000	82,000	361,000
1970	4,087,000	147,000	2,929,000
1971	3,253,000	102,000	2,236,000
1972	3,390,000	119,000	2,251,000
1973	4,239,000	81,000	2,709,000

To estimate the social benefits from the innovation, we must calculate the savings to the consumer arising from the innovation. But this is complicated by the fact that one of the innovator's competitors was working independently on a similar new product, which it introduced very soon after the innovator introduced this new product. The proper comparison here is between what occurred in fact and what would have occurred if the innovator had not innovated, but its competitor did. Based on estimates of the innovating firm, if the innovator had not developed its new product, most of the consumers that benefited from its new product would have bought its competitor's product instead. Thus, the saving due to the innovation is only the *extra* sales of the new kind of product (over what they would have been if the innovator had not introduced its product) times the saving per unit of the innovation.

The innovating firm made estimates of how great these extra sales were each year. Then multiplying these extra sales by the saving per unit of the new type of product over the old, we get the estimates of $(P_1 - P_2)Q_2$ shown in the first column of Table 9.11. Based on the price per use before and after the innovation occurred, it appears that $K = 1.08$; and making the conservative assumption that $n = 1$, we must multiply $(P_1 - P_2)Q_2$ by $(1 - ½ \times 1.08) = 0.46$ to obtain an estimate of consumers' surplus. The deflated value of consumers' surplus is shown in column 2 of Table 9.11.

In addition, it is necessary to take acccount of the fact that competitors would have earned profits on the products displaced by the innovator's new product if the innovation had not occurred. Based on the innovating firm's estimates, the deflated value of these forgone profits were deducted from the last column of Table 9.10, the result being shown in the third column of Table 9.11.

TABLE **9.11**

Social Returns from the Innovation

YEAR	$(P_1 - P_2)Q_2$	DEFLATED CONSUMERS' SURPLUS	INNOVATOR'S DEFLATED NET PROFIT[1] LESS FORGONE DEFLATED PROFITS OF COMPETITORS	DEFLATED ENVIRONMENTAL COSTS	DEFLATED NET SOCIAL BENEFITS
1967	0	0	−$387,000	0	−$387,000
1968	$12,881,000	$4,845,000	−8,926,000	$1,731,000	−5,812,000
1969	32,237,000	11,567,000	180,000	2,964,000	8,783,000
1970	24,673,000	8,389,000	1,465,000	1,989,000	7,865,000
1971	18,090,000	5,877,000	1,118,000	1,163,000	5,832,000
1972	15,654,000	4,929,000	1,126,000	1,210,000	4,845,000
1973	14,928,000	4,462,000	1,355,000	1,351,000	4,466,000

[1] Adjusted for uncommercialized R and D.

This innovation resulted in some environmental costs because its use led to increased water pollution. We asked officials of the Environmental Protection Agency to estimate how much it would cost to remove from the water the amount of chemical that the use of the innovation added to our water supplies. They provided us with a rough estimate of this sort. Based on this estimate, the fourth column of Table 9.11 shows the environmental costs each year. Note that this is a substantial overestimate because, if the innovation had not been introduced, much of this pollution would have occurred as a consequence of increased sale of the competitor's new product.

The deflated net social benefits each year are column 2 plus column 3 minus column 4 of Table 9.11, the result being shown in the last column of Table 9.11. Based on the figures in Tables 9.10 and 9.11, the real internal rates of return are

116 percent (social) and 4 percent (private). These rates of return are conservative since post-1973 benefits are not included. However, it seems doubtful that the private rate of return will ever rise to the point where the innovator will regard the innovation as a commercial success.

7. *Process A*

This is a new design concept applied to a large and complex line of industrial heating equipment. The innovating firm carried out research and development on the innovation in 1959–1961, and the innovation was first used in 1961. In subsequent years, research and development continued to be carried out, as shown in Table 9.12. These figures concerning R and D expenditures were based on detailed data concerning man-hours spent by various kinds of engineering and technical personnel on the project.

The innovating firm received benefits of two kinds from the innovation. First, the innovation reduced its own production costs. It passed some of the cost reduction along to consumers but retained some in the form of extra profits. Second, it licensed the innovation to other firms and received substantial royalties

TABLE **9.12**

Net Private Benefits from Process A

YEAR	GROSS PROFITS	R AND D EXPENDITURES	DEFLATED NET PRIVATE BENEFITS (ADJUSTED FOR UNCOMMERCIALIZED R AND D)
1959	—	$2,000	−$2,000
1960	—	37,000	−48,000
1961	$36,000	144,000	−137,000
1962	168,000	152,000	−12,000
1963	217,000	74,000	142,000
1964	319,000	19,000	320,000
1965	215,000	142,000	51,000
1966	126,000	58,000	60,000
1967	113,000	41,000	65,000
1968	330,000	84,000	222,000
1969	225,000	47,000	155,000
1970	281,000	29,000	212,000
1971	289,000	14,000	225,000
1972	393,000	67,000	251,000
1973	315,000	35,000	206,000

TABLE 9.13

Net Social Benefits from Process A

YEAR	$(P_1 - P_2)Q_2$	CONSUMERS' SURPLUS	DEFLATED NET SOCIAL BENEFITS
1959	—	—	−$2,000
1960	—	—	−48,000
1961	$18,000	$18,000	−117,000
1962	93,000	92,000	89,000
1963	71,000	70,000	218,000
1964	155,000	153,000	484,000
1965	143,000	141,000	200,000
1966	213,000	210,000	276,000
1967	323,000	319,000	384,000
1968	103,000	102,000	361,000
1969	211,000	208,000	344,000
1970	251,000	248,000	426,000
1971	98,000	97,000	305,000
1972	174,000	172,000	389,000
1973	310,000	306,000	436,000

from them. The sum of these two benefits are included in the first column of Table 9.12. To obtain the firm's net profit each year, we must deduct the R and D expenditures in column 2, as well as a prorated share of the uncommercialized R and D costs. After making this calculation, and expressing the result in 1967 dollars, we obtain the results shown in the last column of Table 9.12.

Turning to the social benefits from the innovation, we must calculate the extent of the price decrease to consumers.[6] Based on interviews with the firm, the difference in price of the product due to the innovation times the actual quantity sold equals the amount in the first column of Table 9.13. To estimate the consumers' surplus due to the price reduction, $(P_1 - P_2)Q_2$ must be multiplied by $(1 - \frac{1}{2} Kn)$. According to the available data, K is approximately 0.026. The price elasticity of demand for the equipment is, according to the firm, less than 1. To be on the conservative side, we assumed that $n = 1$, and multiplied $(P_1 - P_2)Q_2$ by $(1 - \frac{1}{2} \times 0.026) = 0.987$. The results are shown in the second column of Table 9.13.

Although two other firms were doing somewhat similar R and D, their results were not transferable in any simple way to this firm, or vice versa. Consequently, it seems appropriate to assume that if this firm had not carried out this innovation, the private and social benefits in Tables 9.12 and 9.13 would not have oc-

6. For definitions of the symbols used in sections 7 and 8 of this chapter, see section 6 of Chapter 8.

curred during this time frame. And the firm's expenditures on R and D (corrected for uncommercialized R and D) are a proper measure of the social costs incurred to obtain the relevant social benefits.

Although this process improvement did displace an old product, there is no reason to adjust the profit figures for the profits that would have been earned on displaced products, since the latter were estimated to be essentially zero. The savings are largely labor savings, and the displaced labor seemed to have no trouble in finding comparable employment elsewhere.

Based on the figures in Tables 9.12 and 9.13, the real internal rates of return are 92 percent (social) and 47 percent (private). These rates of return are conservative since they ignore benefits accruing after 1973.

8. Process B

This innovation is, and has been for the past decade, an important process technology of great benefit to the entire industry of which the innovator is a part. The innovation is based on a series of technical accomplishments. The essence of the technology can, to a great extent, be embodied in special substances that are crucial to the most efficient manner of processing vast amounts of raw material. Largely as a result of these circumstances, numerous firms in an associated supplying industry have managed to manufacture this substance and make it available throughout the industry of the contributing firm. Rivals of the innovating firm have the resources and know-how to simulate the technology associated with using the substances and hence have managed to take full advantage of the whole processing-technology package at little or no cost.

The development of this innovation originated from a sizable evolutionary-type research effort by the innovator, later to become a specific applied R and D program of major importance. Some eleven fairly discrete technical accomplishments have been recorded by the firm which relate to this innovative technology. The first was recorded in 1960, the most recent in 1971, and the rest in a fairly even distribution in number throughout the 1960–1971 period. These accomplishments involved (1) the breakthrough or discovery of the usefulness of these types of substances in the processing operation, (2) significant improvements in substances of this nature, and (3) modifications in processing, equipment utilization, and technique resulting from growing experience with the new substances. A relatively small amount of the total value generated is attributable to accomplishments recorded after 1966.

The program costs in question were very sizable; in several years during the early 1960s they exceeded $2 million annually, more than half of all the applied R and D expenditures by the innovating firm's process group. According to the best recollection of executives of the innovating firm, the program costs began in 1957, although documented program expenses are not available until 1960. The estimated program costs from 1957 to 1959 are simply interpolated amounts based on a zero value in 1956 and the rounded actual expenses identified for 1960. This particular program still exists, although, for purposes of this study, costs are truncated as of 1972 because the compiling of private and social value generated covers only accomplishments recorded through 1972; the truncation will receive further attention in the subsequent discussion of value generated. Program costs include prorated charges by planning and plant general-services groups.

To estimate the ''basic'' research costs associated with this series of related accomplishments, we assume that the maximum percentage of the whole annual applied process R and D group budget that was accounted for by the particular program in question in any given year was the percentage of all process ''basic'' research expense which is attributable to this particular program. Based on this

TABLE **9.14**

Cost of Developing and Implementing Process B

YEAR	''BASIC'' RESEARCH COSTS	APPLIED R AND D COSTS	IMPLEMENTATION COSTS	TOTAL COST TO THE INNOVATOR
			(millions of dollars)	
1953	$0.1	—	—	$0.1
1954	0.2	—	—	0.2
1955	0.3	—	—	0.3
1956	0.5	—	—	0.5
1957	0.5	$0.4	$0.1	1.0
1958	0.6	0.6	0.2	1.4
1959	0.6	1.1	0.2	1.9
1960	0.6	1.2	0.3	2.1
1961	0.7	1.6	0.4	2.7
1962	0.7	2.4	0.5	3.6
1963	0.7	3.2	0.7	4.6
1964	0.7	3.4	0.8	4.9
1965	0.7	2.8	0.6	4.1
1966	0.6	2.4	0.5	3.5
1967	0.5	1.8	0.4	2.7
1968	0.4	1.6	0.3	2.3
1969	0.2	1.6	0.3	2.1
1970	0.2	1.2	0.2	1.6
1971	0.1	1.0	0.2	1.3
1972	—	1.4	0.3	1.7

Note: All R and D and implementation costs are adjusted for uncommercialized projects.

assumption, the fraction of these expenditures (the same fraction applies for each year involved) that is assumed to be relevant to the subset of innovative activity which gave rise to this particular program is 55 percent.

An allowance for implementation costs associated with this technology is made in the same manner as discussed in connection with industrial product C. Also, as in earlier case studies, applied R and D costs and implementation costs used in the construction of investment income streams reflect an appropriate allowance for expenditures on other projects which failed to reach commercialization.

All costs associated with this subset of investment activity are rounded to the nearest hundred thousand dollars and are shown, year by year, in Table 9.14. For simplicity, the whole innovative technology package that is currently being examined, and is made up of numerous technical accomplishments, is henceforth referred to as process B.

The savings to the innovating firm from process B could be estimated quite accurately by the firm. We were fortunate because the firm had already prepared a special engineering report aimed at estimating these savings. This report also estimates the savings to the rest of the industry. As noted above, other firms have been able to use the innovation at little or no cost (for the technology). Table 9.15 shows the resource saving to the innovating firm each year, as well as the total saving in the entire industry. The estimates end at 1972, and we arbitrarily assume that no benefits accrue after that date. This results in a very conservative estimate of both the private and social benefits.

TABLE **9.15**

Gross Private and Social Benefits from Process B

	BENEFITS TO INNOVATOR		SOCIAL BENEFITS	
YEAR	RESOURCE SAVING	EXTRA SAVING OVER SITUATION IF IT HAD NOT INNOVATED	RESOURCE SAVING	EXTRA SAVING OVER SITUATION IF FIRM HAD NOT INNOVATED
		(millions of dollars)		
1962	$4	$4	$4	$4
1963	7	7	10	10
1964	24	24	93	93
1965	25	25	220	220
1966	15	11	297	193
1967	15	8	302	250
1968	19	8	374	286
1969	29	8	433	272
1970	29	10	450	207
1971	26	11	437	156
1972	26	7	419	87

There is evidence that other firms and research organizations were working in this area and that, if the innovator had not come forth with this process, another firm would have. But it is very difficult to pinpoint the exact date when it would have become available. Based on conversations with people in the innovating firm and elsewhere, it appears that a reasonable, but conservative (i.e., short) estimate of the time when it might have been available was four years after the actual date. If this is so, we can approximate the benefits from the process's being available four years earlier by assuming that the diffusion of the new process would have gone on at the same rate as in fact it did, but that the diffusion process would have begun four years later. Thus, we can deduct the actual resource saving in Table 9.15 for a particular year from that four years earlier to estimate the extra saving over what would have occurred if the innovation had occurred four years later.[7] The results of these computations are shown in Table 9.15.

Based on these figures, the stream of net private benefits and net social benefits from the innovator's investment in the innovation—which made it available four years earlier than would otherwise be the case—is shown in the first two columns of Table 9.16. The private benefits are quite conservatively estimated since it is assumed that if the innovator had not innovated, it could have obtained the technology without cost. Also, both the social and private benefits are on the low side because post-1972 benefits are ignored.

Using these rather conservative assumptions, the nominal internal rates of return are 77 percent (social) and 34 percent (private), the corresponding real rates of return being 74 percent and 31 percent. These figures seem quite robust in the sense that one can make quite different assumptions about the extent of the time lag before the innovation would have appeared, and the results vary by only small amounts.

Finally, it is interesting too to compute the social rate of return from the entire industry's investment in R and D directed toward this innovation. According to the innovating firm (and other sources), the total investment by the innovator and its competitors in this area was approximately (and almost certainly no more than) three times that of the innovator. If the investment figures in Table 9.14 are adjusted in this manner, and if it is assumed (quite sensibly) that the innovation would not have come along if none of the relevant firms invested in it, the stream of net social benefits is as shown in the third column of Table 9.16. The nominal internal social rate of return from the industry's total investment is 59 percent, the corresponding real rate of return being 56 percent. This calculation has the advantage that no assumption has to be made about when the innovation would have occurred if the innovator had not invested in it (other than that it would not have occurred before 1973). This is the estimate of the social rate of return that is used in Table 8.1 in the previous chapter.

7. The saving in plant and equipment was not assumed to be postponed four years ahead, since its replacement was probably not postponable.

TABLE **9.16**

Net Social and Private Benefits from Process B

YEAR	PRIVATE	SOCIAL	SOCIAL, BASED ON RETURNS TO ENTIRE INDUSTRY INVESTMENT
		(millions of dollars)	
1953	−$0.1	−$0.1	−$0.3
1954	−0.2	−0.2	−0.6
1955	−0.3	−0.3	−0.9
1956	−0.5	−0.5	−1.5
1957	−1.0	−1.0	−3.0
1958	−1.4	−1.4	−4.3
1959	−1.9	−1.9	−5.9
1960	−2.1	−2.1	−6.4
1961	−2.7	−2.7	−8.2
1962	0.4	0.4	−6.7
1963	2.4	5.4	−3.8
1964	19.1	88.1	78.2
1965	20.9	215.9	207.7
1966	7.5	189.5	286.6
1967	5.3	247.3	293.9
1968	5.7	283.7	367.2
1969	5.9	269.9	426.6
1970	8.4	205.4	445.3
1971	9.7	154.7	433.1
1972	5.3	85.3	413.9

9. *Summary and Conclusions*

In this chapter, we have presented brief summaries of seven of the seventeen case studies. The case studies described here relate to three industrial products, two consumer products, and two processes. Having worked through these case studies, the reader should have a much clearer picture of exactly how the social and private rates of return in Chapter 8 were computed. Also, these case studies should provide a unique description of important aspects of the innovation process. For the first time, data are provided concerning the stream of benefits and costs resulting from a sample of industrial innovations.

The results indicate several important points concerning the innovation process. First, they show the key role of competition in driving a wedge between the private and social returns from innovation. In over half of the cases described in this chapter, competitors began producing (or using, in the case of process B) the innovation within a short interval after the innovation was first commercialized. Because of this competition, there was no way that the innovator could appropriate nearly all of the social returns from the innovation.

Second, the results help to show why, for some types of innovations, the private rate of return to the innovator may exceed the social rate of return from the innovation. As illustrated by our discussion of industrial product A, some innovations enable the innovator to gain a considerable amount of business from its competitors. In cases where the social benefits from such innovations are only moderate, the private rate of return may exceed the social rate of return. Judging from the results of the previous chapter, this situation occurs in a minority, not a majority, of cases. But it is interesting to have detailed data concerning such cases.

Third, the results show that even when an innovator's work results in the faster introduction of the innovation by only a relatively short time interval, the social rate of return from its investment may be very high. For example, in the case of consumer product B, the innovator hastened the innovation's introduction by less than a year. Nonetheless, the social rate of return from the innovator's investment was over 100 percent. Also, in the case of process B, if the innovator hastened the innovation by four years, the social rate of return from its investment was over 70 percent.

Fourth, the results provide some interesting evidence concerning the large social returns to be obtained from relatively mundane innovations. Certainly, thread or household cleaners do not stir the imagination like some other types of products, and scientific and engineering work in such areas may not be as prestigious as in some other areas. But it turns out that the social rate of return from the thread innovation was over 300 percent, and the social rate of return from the cleaning innovation was over 200 percent. Without denying the importance of our "high technology" industries, one is led to wonder whether there may not be a tendency in some quarters to underestimate the returns obtainable from more R and D and related innovative activities in the relatively "low technology" industries.

10 CONCLUSIONS

1. Introduction

The purpose of previous chapters has been to present new econometric findings concerning the social and private rates of return from industrial innovations, the nature of the development process, the determinants of development costs, the relationship between innovation, development, and size of firm, the organizational and strategic factors associated with probabilities of success in industrial research and development, and the determinants of the speed of application of new technology. As indicated in Chapter 1, each of these topics is of central importance, both to analysis and policy in this area; and each is badly in need of further illumination. In this concluding chapter, we bring together our findings and discuss some of their implications.

2. Social Returns from Investments in New Technology

From the point of view of public policy, perhaps our most interesting results pertain to the social rates of return from investments in new technology. Although it has long been recognized by economists that estimates of such rates of return were of crucial importance in formulating any rational policy toward civilian technology, no estimates of this sort have been made for industrial innovations.[1] In Chapter 8, we presented a model which indicates how such estimates can be made for many, but by no means all, innovations. This model includes the pricing behavior of the innovator, the effects of the innovation on displaced products, and the costs of uncommercialized R and D and of R and D outside the innovating organization, as well as the factors contained in earlier models of this sort.

1. Only in the case of a few nonindustrial innovations have such estimates been made. See E. Mansfield, "Federal Support of R and D Activities in the Private Sector" in *Priorities and Efficiency in Federal Research and Development* (Washington, D.C.: Joint Economic Committee of Congress, 1976), for a discussion of previous estimates.

We obtained very detailed data concerning 17 innovations from the producers and users of the new technologies, and applied this model to obtain an estimate of the social rate of return in each case. Practically all of these innovations were of average or routine importance, not major breakthroughs. (For one thing, we wanted to avoid biasing the sample toward innovations that probably had relatively high rates of return.) Although the sample cannot be regarded as randomly chosen, there is no obvious indication that it is biased toward relatively profitable innovations (socially or privately) or relatively unprofitable ones. In large part, it contains undramatic, run-of-the-mill improvements in products and processes, like a new type of thread or an improved machine tool. As we pointed out in Chapter 1, this is the type of work that accounts for most industrial research and development.

The results are quite striking. The median social rate of return from these 17 innovations was about 56 percent, which indicates that the investments in these new technologies paid off handsomely from society's point of view. To check this result, we used the same model to estimate, for one of the nation's largest firms, a lower bound on the social rate of return from its investment in new process technology in 1960–1972. This result too was about 50 percent. Of course, our sample, although large relative to what was previously available, is too small to support definitive conclusions, but the results certainly suggest that, even taking into account the riskiness of innovative activity, the rate of return from investments in new technology has tended to be high.

Economists are interested in estimates of social rates of return because they provide clues as to whether we as a nation are under-investing or over-investing in civilian technology. If the marginal social rate of return from investments in civilian technology is greater (less) than the marginal social rate of return from other uses of the relevant resources, this is evidence of an under-investment (over-investment) in civilian technology. Unfortunately, our results pertain to the average, not the marginal social rate of return from investments in civilian technology. William Fellner and Zvi Griliches have argued that it is legitimate—or at least not too rash—to make the jump from average to marginal rates of return.[2] Indeed, in Griliches's view, there is no reason to believe that the marginal rate of return differs much from the average rate of return. If this is the case, our results certainly suggest that there may be an under-investment in civilian technology in the United States, since the average rate of return seems very high.

Our finding concerning rates of return is quite consistent with other econometric studies that have relied on more indirect methods and have used more highly aggregated data. In a previous study based on statistical production functions, one of the present authors found that the marginal rate of return from

2. W. Fellner, "The Progress-Generating Sector's Claim to High Priority," and Z. Griliches, "A Memorandum on Research and Growth," both in *Research and Development and Economic Growth/Productivity* (Washington, D.C.: National Science Foundation, 1972).

R and D in the chemical and petroleum industries was 30–40 percent.[3] Minasian[4] obtained similar results. In a more recent study, Terleckyj[5] has estimated about a 30 percent rate of return from an industry's R and D (in manufacturing), based only on the effects of an industry's R and D on its own productivity. In addition, his findings suggest a very substantial effect of an industry's R and D on productivity growth in other industries, resulting in a social rate of return greatly exceeding the 30 percent figure. Without exception, econometric studies based on production functions point in the same direction as our studies in Chapters 8 and 9.[6] Although these and other investigations (including our own) have noteworthy limitations, it is remarkable that the results, which rely on quite different methods and data, are so similar.

3. Social versus Private Rates of Return

To understand why there may be some under-investment in civilian technology, one must recognize that private rates of return may not equal social rates of return. As is evident from our 17 case studies, as well as the detailed investigation of the R and D activities of the major firm (in Chapter 8) during 1960–1972, firms often cannot appropriate all of the social benefits from an innovation. A good example is the new type of thread (industrial product B) described in Chapter 9. Although the social rate of return was over 300 percent, the private rate of return was only 27 percent, partly because other firms began imitating the new thread within six months after the innovator introduced it.

Our results provide the first glimpse of the differences between private and social rates of return from investments in new technology. In general, social rates of return seem to exceed private rates of return, although this is not always the case. Specifically, the median social rate of return is about double the median private rate of return in our sample, and the social rate of return from the major firm's investment in new process technology in 1960–1972 seemed to be at least double the private rate of return. When we look at specific innovations, the difference between the social and private rate of return seems to be related to the economic importance of the innovation (measured by absolute annual benefits)

3. E. Mansfield, *Industrial Research and Technological Innovation* (New York: Norton for the Cowles Foundation for Research in Economics at Yale University, 1968).

4. J. Minasian, "Research and Development, Production Functions, and Rates of Return," *American Economic Review* (May, 1969).

5. N. Terleckyj, *Effects of R and D on the Productivity Growth of Industries: An Exploratory Study* (Washington, D.C.: National Planning Association, 1974).

6. E. Mansfield, "Federal Support of R and D Activities in the Private Sector," *op. cit.*

and to the costs of imitating the innovation. These results are quite consistent with hypotheses put forth by Arrow,[7] Matthews,[8] and others.

Socially worthwhile innovations—that is, projects where the social rate of return is high enough to warrant going ahead with the project—will not be carried out if the perceived private rate of return is so low that the potential innovator rejects the project. An important question facing policy makers is: how frequently does this situation arise, and in what areas is it most prevalent? If this situation is common, it suggests the desirability of direct or indirect government support for such projects. If it is much more common in some types of industries and for some types of innovations than for others, then this support should be focused on such industries and such types of innovations. Economists have long recognized that this question lies at the heart of any discussion of public policy toward civilian technology,[9] but until now there has been no direct evidence concerning it.

Our results cannot indicate a great deal about the frequency with which such situations arise, because our data pertain to innovations that were carried out. (The difficulty, if not impossibility, of obtaining such data for innovations that were not carried out should be obvious.) But a limited amount of evidence can perhaps be derived on this score from the data presented in Chapter 8 concerning the 17 innovations. For 9 of these innovations, we could obtain data concerning the approximate private rate of return *expected from the innovation* by the innovator when it began the project. In 5 of the 9 cases, this expected private rate of return was less than 15 percent (before taxes), which indicates that these 5 projects were quite marginal from the point of view of the firm. (Indeed, the executives of the firms confirmed that they were marginal.) Yet *the average social rate of return of these 5 innovations was over 100 percent.*

Unless the social rate of return drops precipitously when the expected private rate of return falls from 10 or 12 percent to (say) 5 or 6 percent, this result seems to indicate that there may be many projects where the expected rate of return was a bit lower than for these 5 projects (with the result that they were not carried out), but where the social rate of return would have been quite high. Among the innovations for which we have data, there is no significant correlation between an innovation's expected private rate of return and its social rate of return. Thus,

7. K. Arrow, "Economic Welfare and the Allocation of Resources for Invention," in National Bureau of Economic Research, *The Rate and Direction of Inventive Activity* (Princeton: Princeton University Press, 1962).

8. R. C. O. Matthews, "The Contribution of Science and Technology to Economic Development," in B. Williams, ed., *Science and Technology in Economic Growth* (London: Macmillan, 1973).

9. For example, see R. Nelson, M. Peck, and E. Kalachek, *Technology, Economic Growth, and Public Policy* (Washington, D.C.: Brookings Institution, 1967).

unless there is a sharp discontinuity in the slope of whatever relationship exists between the expected private rate of return and the social rate of return, no evidence exists to suggest a precipitous drop in the social rate of return when the expected private rate of return falls from 10 to 12 percent to 5 or 6 percent. These results, like those discussed in the previous section, may point toward some under-investment in civilian technology.

4. Public Policy Alternatives

At least three kinds of policy mechanisms have been suggested to help deal with whatever under-investment may exist in civilian technology—government grants and contracts to industry and universities for more such work, increased use of government laboratories for such purposes, and tax credits for private industry.[10] Our results indicate some of the problems in each of these mechanisms. With regard to grants and contracts, as with any selective mechanism, one runs into the problem that benefits and costs of various kinds of R and D are very hard to forecast. As we have seen, even major corporations have difficulties using various forms of cost-benefit analysis for R and D project selection, although they have a benefit concept that is much easier to estimate than most government agencies do. To illustrate how far off benefit estimates are, the chances were about 50-50 that a new product's (or new process's) estimated discounted profits would be more than double, or less than one-half, the actual discounted profits in the major firm studied in Chapter 8. Further, as Eads and others have stressed,[11] such estimates may be biased for parochial, selfish, or political reasons, the result being a distortion of social priorities, if the estimates are taken seriously.[12]

10. For example, see the testimony by R. Nelson and E. Mansfield before the House Subcommittee on Science, Research and Development in *Science, Technology and Economic Growth* (Washington, D.C., Government Printing Office, 1971).

11. G. Eads, "U.S. Government Support for Civilian Technology: Economic Theory vs. Political Practice," *Research Policy,* 1974. For some results in an industrial context, see E. Mansfield, J. Rapoport, J. Schnee, S. Wagner, and M. Hamburger, *Research and Innovation in the Modern Corporation* (New York: Norton, 1971).

12. These are not, of course, the only problems with the use of government grants and contracts. According to some studies, although an industry's own R and D expenditures have a significant effect on its rate of productivity increase, the amount of federally-financed R and D performed by the industry seems to have little or no such effect. See Terleckyj, *op. cit.* In part, this may be due to the possibility that output measures in industries like aircraft are not reliable measures of social value. But it may also be due to a difference in the effectiveness of federally financed and privately financed R and D. More work needs to be done regarding this question. For some relevant discussion, see K. Pavitt and W. Walker, "Government Policies toward Industrial Innovation," University of Sussex, 1974.

With regard to the increased use of government laboratories to promote civilian technology, our results in Chapter 2 suggest the problems in having R and D conducted by organizations that are not in close touch with the marketing and production of the product. It is very important that there be unimpeded flows of information and good coordination of R and D, on the one hand, and marketing and production, on the other. Otherwise, the R and D is likely to be misdirected, or even if it is not, it may be neglected or resisted by potential users. As we saw in Chapter 2, this is a difficult problem for various divisions of a firm, and it would seem to be made worse if the R and D is done in government laboratories. In the last decade, many governments have tended to convert government laboratories and to increase the amount of government-financed R and D done in industrial firms in order to bring R and D into closer contact with application and commercialization.

With regard to tax credits, it is evident that they would reward firms for doing R and D that they would have done anyway, that they would not help firms with no profits, and that they would encourage firms to define R and D as widely as possible. Some of these problems might be solved by a tax credit for increases in R and D, but many problems remain. For example, firms would still have an incentive to redefine R and D, and for firms that can appropriate little of the social returns from new technologies, R and D would still be unprofitable even if the tax credit existed.

At present, it is difficult, if not impossible, to specify what combination of these (and other) mechanisms would be most effective in compensating for whatever under-investment there is in civilian technology. But it seems likely that any such civilian technology program should be neither large-scale nor organized on a crash basis, that it should not focus on helping sick industries (merely because they are sick), that it should not get the government involved in the latter stages of development work, that a proper coupling should be maintained between technology and the market, and that the advantages of pluralism and decentralized decision making should be recognized. Given the current uncertainties (which, as we have repeatedly stressed, are great), it wouls seem wise to proceed with considerable caution, and to build into any program the capacity and necessity to resolve many of the key uncertainties before too big a commitment is made.[13]

5. *Private Rates of Return*

To understand the process of technological change, it is important to obtain information concerning the private rates of return, as well as the social rates of return, from investments in new technology. Our results provide some of the first

13. For further discussion and amplification of these points, see E. Mansfield, "Federal Support of R and D Activities in the Private Sector," *op. cit.*

published direct estimates of such private rates of return. For the 17 innovations in Chapter 8, the median estimated private rate of return was about 25 percent. For the large industrial firm, the estimated private rate of return from its investment in new process and product technology in 1960–1972 was about 19 percent. In both cases, the rates of return are before taxes. The riskiness of investments of this kind is evidenced by the enormous variation among the private rates of return for the 17 innovations. For 6 of these innovations, the private rate of return was less than 10 percent, while for 5 of them, it was more than 40 percent.

Private rates of return are important because they determine how much firms invest in innovative activities. As pointed out in an earlier chapter, industry's total annual R and D expenditure increased appreciably between 1960 and 1969, but declined somewhat in real terms between 1969 and 1974. One possible reason for this levelling-off of industrial R and D expenditures was that the profitability of such expenditures tended to stabilize, or even to fall off during the late 1960s and early 1970s. With this in mind, we find it interesting that our results in Chapter 8 suggest that, for the major firm studied there, the private rate of return during the late 1960s and early 1970s tended to be lower than during the early 1960s. We suspect that this was the case for many other firms as well.[14]

A firm's private rate of return from its innovative activities is related to the productivity of its research and development, among other things. In Chapter 2, we presented data (apparently for the first time) concerning interfirm differences in the proportion of R and D expenditures that goes for economically successful projects.[15] So long as there is little or no correlation between the size of this proportion and the average rate of return from the firm's economically successful projects, this proportion should be a useful, if rough, measure of the productivity of a firm's R and D. According to the available data, this condition is satisfied, at least for the firms in our sample. Our results indicate that there are very substantial differences among firms in these proportions, or probabilities of economic success. Even in the same industry, these probabilities can be much bigger in one firm than in another. For example, in three reasonably comparable chemical firms, the proportions of R and D expenditures that were economically successful in 1968–1971 were 69 percent, 54 percent, and 39 percent. These differences seem much too large to be attributed to errors of measurement or differences among firms in the interpretation or definitions of terms.

14. For some estimates of private rates of return based on statistical estimation of production functions from data for almost 900 firms, see Z. Griliches, "Returns to Research and Development Expenditures in the Private Sector," Conference on Research in Income and Wealth, 1975.

15. A project is defined as economically successful if it yields a rate of return (on the R and D costs plus any additional investment made to introduce the innovation) in excess of what was available from other (non–R and D) investment alternatives such as expansion of facilities to produce existing products.

6. Factors Associated With Probabilities of Success

A firm's probability of economic success is the product of three probabilities—the probability of technical success, the probability of commercialization (given technical success), and the probability of economic success (given commercialization). According to the econometric results in Chapter 2, which are based on data from 20 firms in the chemical, drug, petroleum, and electronics industries, all three of these probabilities of success are directly related to how quickly R and D projects are evaluated from the point of view of economic potential. Apparently, this reflects the fact that firms sometimes allow projects to run on too long before evaluating their economic (as distinct from their technological) potential. Even though it is very difficult to forecast the benefits and costs of a particular R and D project, firms seem to do better if they try to do so early in the game, rather than wait until substantial amounts have been spent.

In addition, our results indicate that the probability of commercialization (given technical completion) is directly related to the degree to which R and D and marketing are integrated. It seems likely that this reflects the fact that, in some firms, the R and D department has not always worked very closely with the marketing staff, the result being that the R and D output has not been as well mated with market realities as it might have been. In previous studies, we found indications of a similar tendency, but the results presented here are much stronger than those we obtained before.[16]

Further, our findings suggest that a substantial percentage of a firm's R and D results may lie fallow because other parts of the firm do not make proper use of them. According to estimates made by executives of 18 of these firms, the percentage of R and D projects that were economic successes would have increased by about one-half if the marketing and production people had done a proper job in exploiting them. And it is important to note that the non-R and D executives seemed to agree on this point with the R and D executives. If this figure is anywhere close to the truth, it suggests that faulty interfaces between R and D and the rest of the firm result in a very substantial decrease in the productivity of industrial R and D.

Of course, firms have tried a variety of devices and strategies to promote better utilization of their R and D findings. For example, they have moved personnel back and forth across the interface between the R and D department and other departments, and they have established new product committees and new product departments. It seems likely that these devices have had a beneficial effect. For example, our results suggest that R and D results tend to be more fully utilized in firms with new product committees. But much more needs to be done, since the problem of "X-inefficiency" seems very important in this area. If such inefficiency could be reduced, there is apparently the opportunity for a substan-

16. E. Mansfield *et al., op. cit.*

tially higher rate of technological change without any increase in R and D expenditures.

7. Tradeoffs Among Probabilities and the Choice of an R and D Policy

Our results also indicate that a tradeoff exists among some of these probabilities of success. For example, there is a tradeoff between the probability of technical completion and the probability of economic success (given commercialization). As more of the R and D portfolio stems from the R and D department (which tends to be the source of technically more ambitious and fundamental projects), the probability of technical completion goes down, but the probability of economic success (given commercialization) goes up. This seems reasonable because technically more ambitious projects would be expected to have a smaller chance of being technically completed; but if completed (both technically and commercially), they would be expected to have a greater chance of being economically successful.

Firms must recognize that this tradeoff exists, and decide on rational grounds how far they want to go in reducing one probability to increase the other. At present, some firms seem to keep the probability of technical completion at a high level, even though a reduction in this probability might be beneficial because it might result in a substantial increase in the probability of economic success (given commercialization). In other words, some firms carry out many relatively unambitious projects where the probability of technical completion is high, but where the potential profit is low. They might do better to devote a bigger share of their R and D expenditures to technically more ambitious projects where the probability of technical completion is lower but where the probability of economic success (given commercialization) is higher.[17]

Also, firms should recognize that there is a tradeoff between the probability of commercialization (given technical completion) and the probability of economic success (given commercialization). For example, quantitative project selection techniques may increase the probability of commercialization (given technical completion), since they force a more careful evaluation of each project's commercial potential. On the other hand, they may reduce the probability of economic success (given commercialization), since they may push a firm's R and D portfolio in the direction of less ambitious projects.[18] Whether such a

17. Of course, it is possible to go too far in the other direction too. For some interesting discussion, see E. Gee and C. Tyler, *Managing Innovation* (New York: John Wiley, 1976).

18. For an interesting discussion of project selection techniques (and many other topics), see C. Freeman, *The Economics of Industrial Innovation* (Baltimore: Penguin, 1974).

tradeoff is beneficial or not will vary from firm to firm. But it is important that firms recognize that such tradeoffs exist, and that they attempt to respond rationally to them.

8. Analyzing the Development Process

The development process is at the heart of the production of new industrial technology. But because it is a highly technical process that is not well defined and that varies from industry to industry, economists and policy analysts have not understood it well enough to model it satisfactorily. As a first step toward the construction of such a model, it is necessary to categorize the various activities that are part of development, and to determine the purposes of each activity, the inputs it uses, the outputs it produces, the input-output relationships for this activity, and the ways that various activities intermesh and interact. For a few industries, some information of this sort is available. (For example, Marschak, Klein, Peck, and Scherer[19] have provided considerable data for the aerospace industry.) But such studies are extremely few and far between, because economists generally have avoided much entanglement with the details of the development process.

In this book, we have provided a rather detailed and intensive look at the nature of the development process for polymers. Our findings indicate that the development process—defined somewhat more broadly than by NSF[20]—can be decomposed into a number of reasonably well-defined activities, and that data can be obtained from firms regarding each of these activities. Based on data for 56 development projects, it turned out that, on the average, about 93 percent of the total development costs were spent on four activities—pilot plant work, applications research, market development, and bench scale work. Specifically, about 50 percent of the total costs went for pilot plant work, about 19 percent for applications research, about 14 percent for market development, and about 12 percent for bench scale work.[21]

19. T. Marschak, T. Glennan, and R. Summers, *Strategy for R and D* (Springer Verlag, 1967); B. Klein, "Policy Issues Involved in the Conduct of Military Development Programs," *The Economics of Research and Development* (Columbus: Ohio State, 1965); and M. Peck and F. Scherer, *The Weapons Acquisition Process* (Cambridge: Harvard, 1962).

20. Although the definition of development is somewhat more inclusive than the NSF definition, our results in Chapters 4 and 5 can easily be modified to correspond to the NSF definition.

21. Many of the projects in the sample were carried out a number of years ago. Because of changes over time in the development process, these percentages might be somewhat different now. For example, some chemical engineers assert that the pilot plant work has become less important in chemical development.

Averages of this sort provide an overall description of the relative importance—in terms of the value of resources used—of various activities, but they tell us nothing about the way in which the relative importance of these activities changes during the course of a project. According to our data, there tends to be considerable overlap among these activities, many of them being carried out at the same time. In the early stages of a development project, the bench scale work tended to be the dominant activity. Frequently, the major portion of the bench scale work occurred before most of the pilot plant work and market development. During the last half of a project, the pilot plant work tended to absorb more resources than the other activities. The pilot plant work often overlapped the market development. For those projects where a market development effort is required, it is quite understandable that the relationship between the pilot plant work and the market development will be close, since the pilot plant often produces material to be used in market development.

Perhaps the most notable characteristic of development projects is their diversity. Thus, as one would expect, there was a great deal of variation among projects in the relative importance of various activities. Much of this variation was due to differences in the purposes of the development projects. For example, projects aimed at entirely new products generally included all of these activities, whereas projects aimed at product improvements generally included no market development (and often no applications research as well). On the other hand, projects aimed at entirely new processes generally included all of these activities other than market development, whereas projects aimed at process improvements generally included no applications research (and frequently no bench scale work either), as well as no market development.

9. Forecasting Development Costs

This descriptive and classificatory work is useful, indeed essential, because there is not much chance of modeling the development process successfully unless one knows something about its characteristics. However, once we have these descriptive results, the next task is to use them to devise techniques to help answer some of the important problems in this area. In this book, we have used this classificatory scheme to build an econometric model to forecast the development costs for a project. The problem of forecasting development costs is generally agreed to be a central one in this area. Earlier work[22] has shown that the forecasts made by firms (and government agencies) of development costs contain such large errors that they may be of limited use in rational decision-making. Earlier work[23] has also suggested that relatively simple econometric models can out-per-

22. Mansfield et al. *op. cit.*

23. *Ibid.*

form the estimates made by the firms, at least in the cases studied. The question addressed in this book is: can a model that disaggregates the development process into the activities discussed above do better than one that does not disaggregate it in this way?

An equation was estimated to explain the cost of each of the four major activities cited above. Moreover, equations were estimated to explain some of the variables used in each of these four equations, since the model is recursive in form. Together, this system of equations can be used to forecast the total development cost of a project. Judging from five such forecasts, the results seem encouraging. The model's forecasts were more accurate than those of the firms, the root-mean-square-deviation-from-1 of the ratio of actual to estimated costs being 0.60 for the model's forecasts and 1.00 for the firms' forecasts. And more important, the relative improvement over the firms' forecasts was greater, on the average, than in previous studies where the development process was not disaggregated.[24] Thus, based on this crude comparison, our simple disaggregation procedure seems to have improved our forecasting ability. Of course, this does not mean that this model is anything more than a crude aproximation, but it does suggest that, if one is serious about forecasting development costs, it is often worthwhile to disaggregate to the level of these individual activities comprising the development process. For many purposes, the extra precision seems likely to be great enough to offset the extra costs of model construction.

10. "Learning by Doing" in R and D, or "Learning to Learn"

Economists are well acquainted with the process of "learning by doing." Arrow[25] and others have emphasized the importance of this process by which small improvements occur in a production unit—a plant or firm—as its cumulated output grows. However, although the concept of learning by doing is well accepted with regard to production, economists have neglected the fact that similar forms of organizational learning occur in R and D. In this book, we have attempted to investigate the extent of learning by doing in R and D, this investiga-

24. Using a more aggregated approach, the root-mean-square-deviation-from-1 of the ratio of actual to estimated development cost was about 30 percent less for the model-based forecasts than for the firm's forecasts. This is an unweighted average of the results for the three types of innovations considered in *ibid*, p. 84. In contrast, the root-mean-square-deviation-from-1 of the ratio of actual to estimated development costs is about 40 percent less for the model-based forecasts than for the firm's forecasts in the cases discussed above.

25. See K. Arrow, "The Economics Implications of Learning by Doing," *Review of Economic Studies*, 1962; and W. Hirsch, "Manufacturing Progress Functions," *Review of Economics and Statistics*, 1956.

tion being of two types. First, in our econometric model of the determinants of development costs in polymers, we studied the effect of a firm's R and D experience on its development costs. Based on this model, it appears that, all other things equal, the development cost for a polymer seems to decrease, on the average, by about one and one third percent for every year of additional R and D experience in the area on the part of the firm carrying out the development. Clearly, learning by doing seems to be a quantitatively important force in R and D, at least in this sample of projects.

Second, in our study of the factors associated with the differences among innovations in the percent of total innovation cost devoted to R and D, we included the firm's R and D experience as one of the explanatory factors. This variable turns out to have a statistically significant effect (at the .10 level); the percentage of total innovation costs devoted to R and D tends to decrease as the firm gains more and more R and D experience. This would be expected, since the learning in non-R and D aspects of the innovation process is probably not associated strongly with how long the firm has been doing R and D in this area. Judging from our results, a decade of additional experience reduces the percent of total innovation costs devoted to R and D by about 6 percentage points, which again is a quantitatively significant amount.

Thus, both types of investigations yield statistically significant and quantitatively impressive evidence of learning by doing in R and D. We believe that this phenomenon—learning by doing in R and D—plays an important and pervasive role in many areas of economic behavior. To understand why some organizations in the United States are more efficient at particular types of development work than others, one must take proper account of the differences in the extent and nature of their experience. And to understand why some R and D organizations have few competitors, one must take proper account of the barriers to entry erected by the existing organizations' long experience in the area. Even in the People's Republic of China, there is evidence that learning by doing has been an important reason why the older firms in Shanghai have proved more innovative than the new enterprises created with Soviet help.[26]

One reason why learning by doing in R and D has not received much attention from economists is that their basic analytical tools are not equipped to encompass it. The production function, which is the economist's basic analytical device to represent the state of technology, simply classifies input-output combinations as being possible or impossible, whereas the real question for many purposes is how much money (and time) it is likely to take to make them possible. One way to handle such problems is to generalize the concept of the production function. Rather than think in terms of input-output combinations that are "pos-

26. T. Rawski, "Problems of Technology Absorption in Chinese Industry," *American Economic Review*, May 1975. Note that we are not asserting that more experience continues indefinitely to have a positive effect. Beyond some point, organizations may tend to become rigid and stodgy.

FIGURE **10.1**

Isorand Curves

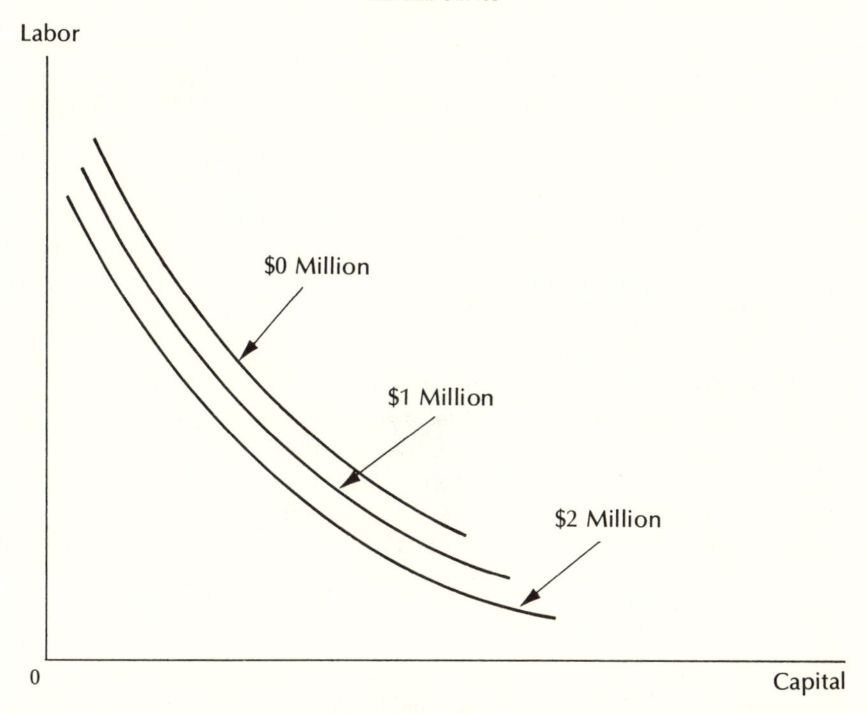

sible" or "impossible," we can think in terms of the cost involved in attaining various input-output combinations. For example, as shown in Figure 10.1, we might draw various "isorand" curves, each of which would indicate the input-output combinations that will be available, given that a particular expenditure is made by the firm during the relevant period on R and D and related activities.[27] For example, the $1 million "isorand" curve in Figure 10.1 shows the various combinations of inputs that could be used to produce 1 million tons of steel per year if $1 million were spent on R and D and related activities. Similarly, the $2 million isorand curve shows the various combinations that could be used if $2 million were spent on R and D and related activities. And the isorand curve corresponding to a zero expenditure on R and D is, of course, the isoquant of traditional theory.

Once we think in terms of these isorand curves, it is a simple matter to represent learning by doing in R and D. The essential point is that a given isorand

27. For simplicity, uncertainty is ignored here. Clearly, it could (and should) be introduced in more extensive discussions. However, it is not essential for present purposes. Alvin Harman of the RAND Corporation has been engaged in work directed at the measurement of relationships like these curves.

curve will tend to be lower and farther to the left for firms with much R and D experience in the relevant area than for those with little experience. For example, if the firm in Figure 10.1 had more R and D experience, its $1 million isorand curve (and its other such curves) would be lower and farther to the left than it is now. Perhaps this simple graphical exercise will help to link the concept of learning by doing in R and D—or of "learning to learn"—to the conventional theory of production and thus make it fit better into place for economists. We hope so, because our findings indicate that the phenomenon of "learning to learn" is important and worthy of further attention.

11. The Role of Large Firms in the Innovation Process: The Case of Chemicals

One of the central issues of antitrust policy (and microeconomic policy, in general) is concerned with the role of the biggest firms in the innovation process. Galbraith, Schumpeter, and others have argued that the very largest firms tend to carry out a disproportionately large share of the major innovations.[28] According to previous studies that we have carried out, this argument is frequently at odds with the facts. In steel, the firms that seem to carry out the most innovations relative to their size, have been relatively small ones. In petroleum and coal, they seem to have been about the sixth-largest firms; and in drugs (at least up to 1962), about the twelfth-largest firms. In this book, we have looked in some detail at the innovative performance of firms of various sizes in the chemical industry. In the case of process innovations (and developments), there is no evidence that the biggest chemical firms did any more innovating (or developing), relative to their size, than somewhat smaller firms. But with regard to product innovations (and developments), the biggest firm, du Pont, turned in a very impressive performance. Relative to its size, it was responsible for the most product innovations (and developments).

Thus, du Pont seems to be the only case encountered thus far where the biggest firm in an industry has done the most innovating (relative to its size)—and even in the case of du Pont, this is true only for products, not processes. This result is in close accord with previous findings concerning the chemical industry, which indicate that this is the one major industry where the biggest firm spends more, relative to its size, on R and D, than somewhat smaller firms.[29] It is also in accord with our previous findings that du Pont has been regarded by other firms

28. J. K. Galbraith, *American Capitalism* (Boston: Houghton Mifflin, 1952); and J. Schumpeter, *Capitalism, Socialism, and Democracy* (New York: Harper, 1950).

29. E. Mansfield, *Industrial Research and Technological Innovation, op. cit.* and F. M. Scherer, "Firm Size, Market Structure, Opportunity, and the Output of Patented Inventions," *American Economic Review,* 1965.

in the industry, and knowledgeable observers outside the industry, as having the most productive R and D establishment in the industry.[30]

One reason why the chemical industry is the single case where the biggest firm has done the most innovating (relative to its size) is that innovations, and particularly new products, have tended to be very expensive in this industry, relative to the distribution of firm sizes. For example, the cost of the major synthetic fiber innovations were not too different from the cost of the major cracking innovations in oil, even though the chemical firms tended to be much smaller than the oil firms.[31] As one of the authors has pointed out elsewhere, the largest firms would be expected to account for a disproportionately large share of the innovations if the advances in technology are such that innovations are very costly relative to the average size of the relevant firms.[32]

But du Pont's innovative performance cannot be attributed to its size alone. Indeed, in many respects, its size may have been a hindrance rather than a help. Other chemical firms that were close to du Pont's size at the beginning of the earlier period that we studied produced few major innovations, while some relatively small firms—like Dow, which was very small in pre-World War II days—did a great deal of innovating (and developing) relative to their size. If du Pont is omitted, the other firms among the biggest four in the industry generally accounted for no larger share of the innovations than they did of the industry's assets. It seems likely that much of du Pont's success during the period we studied was due to the attitudes of its management and to the quality of its personnel, not to its size *per se*.

Certainly, one cannot conclude from these results that the biggest firms—or high levels of concentration—are needed in most parts of the economy to promote rapid technological change. Instead, what they seem to indicate is that, in those industries where the development and introduction of new products and processes require great resources, the largest firms are likely to do a disproportionately large share of the innovating. (Also, they indicate that, just as some small and medium-sized firms have been fortunate and adept in their choice of

30. Mansfield et al., *op. cit.* Of course, these results pertain to the past. Some observers claim that du Pont's lead over its rivals is much smaller now.

31. Simon found that the average investment required to build a plant to produce 29 of the product innovations in Chapter 3 was about $8 million—and this excludes the R and D costs, which averaged about $5 million for some of the major synthetic fibers like nylon, orlon, and dacron [according to J. Jewkes, D. Sawers, and R. Stillerman, *The Sources of Invention* (New York: Norton, 1969)]. Relative to the assets of the major chemical firms at the middle of the relevant period, these costs are greater than were the average costs of the innovations in the steel and petroleum industries, relative to the assets of the major firms in these industries. See E. Mansfield, *Industrial Research and Technological Innovation, op. cit.*

32. Mansfield, *ibid.* Also, see F. M. Scherer, "Antitrust and Patent Policies," paper presented at BMFT-NSF Seminar on Technological Innovation, April 8, 1976.

people and their styles of management, so too have some very large firms, like du Pont.) But it is by no means clear that the costs of innovating are so high in most other industries that very large firms are needed. On the contrary, as noted above, the available data seem to indicate that the chemical industry is the exception, not the rule.[33]

12. The Diffusion Process and Forecasting

After an innovation is introduced, its use must spread in order for its potential economic effect to be achieved. This diffusion process is of enormous importance, both from the point of view of private decision-making and of public policy. About 15 years ago, one of the authors proposed a simple econometric model to represent the diffusion of a technological innovation.[34] This model predicts that the growth over time in the number of firms using a new process should conform to a logistic function, and that the parameter of this function that governs the imitation rate should be related to the profitability of the innovation and the investment required to use it. This model fit his data for 12 innovations in four industries quite well, and subsequent studies by him,[35] Hsia,[36] and others have indicated that it is quite useful for other innovations here and in other countries.[37]

In this book, we have reported three additional batteries of tests to see how well this model performs. First, we described the results of a study of 23 process innovations in the chemical industry, which investigated how well this model can represent the growth over time in the percent of industry output produced with a new process. The results suggest that the model, which can be reinterpreted in these terms, can be used for this purpose. Second, we investigated how well this model can explain the increase over time in the percent of new machine tools purchased that have numerical controls. This is a different measure of the rate of diffusion than the other two measures previously cited—the increase over time in the percent of firms using the innovation, and the increase over time in the per-

33. In their review of the literature, Kamien and Schwartz stress too that the chemical industry is atypical in this regard. See M. Kamien and N. Schwartz, "Market Structure and Innovation: A Survey," *Journal of Economic Literature,* March 1975.

34. E. Mansfield, "Technical Change and the Rate of Imitation," *Econometrica,* October 1961.

35. Mansfield et al., *op. cit.*

36. R. Hsia, "Technological Change in the Industrial Growth of Hong Kong," in B. Williams (ed., *Science and Technology in Economic Growth* (New York: Macmillan, 1973).

37. Also, see L. Nabseth and G. Ray (eds.), *The Diffusion of New Industrial Processes* (Cambridge: Cambridge University Press, 1974).

cent of output produced with the innovation. Our results indicate that this model also can explain the observed differences among industries in this measure of the rate of diffusion of numerically controlled machine tools. Thus, our tests suggest that the model can be used to explain all three measures of the rate of diffusion, a noteworthy result since each of these measures is of importance.

Our third test of this model is somewhat different from the other two. Rather than seeing how well it can fit historical data, we see how well it can forecast under real-life conditions. Some years ago, the Small Business Administration wanted to forecast the rate of diffusion of numerically controlled machine tools in the tool and die industry. Our model, among others, was used to make such forecasts. In this book, we compare the accuracy of this model's forecasts with that of the forecasts of machine tool builders, that of the forecasts based on the plans of the tool and die firms, and that of the forecasts based on the usual naive models. The results suggest that our model yielded the most accurate forecasts.

Clearly, we do not mean to imply that this model is anything more than a crude device, or that it can be applied in situations where its basic assumptions do not hold. But used with caution, it may be helpful in a variety of circumstances. To illustrate how it can be used by managers, consider a firm that is trying to decide how much to charge for a new product which will be used by another industry. Obviously the higher the price the firm charges, the smaller the equilibrium amount that it will sell per year. But this is not the only effect of the price that is charged. In addition, the higher the price, the lower the profitability of the new product to the potential users—and the lower the rate of diffusion of the new product.[38] Our model can be used to estimate the effects of changes in price on the rate of growth of the new product's sales. Based on this model, the (long-run) demand curve for the new product, and its cost function, the firm can estimate which price will maximize its discounted expected profits, once one takes account of the effect of price on both the ultimate sales level and the rate of diffusion.

In addition, of course, this model may prove useful to technological forecasters interested in predicting the rate of diffusion of particular innovations. Some firms are already experimenting with this model for such purposes,[39] and government agencies like the Small Business Administration seem to have found it useful in this regard. But although this model seems to have found useful application, it is far from being a completely adequate representation of the diffusion process, one problem being that it does not include some factors that may

38. In addition, a higher price means a bigger investment by the users, which may also slow up the diffusion process. But for simplicity, we shall ignore this effect here.

39. For example, see A. W. Blackman, "The Rate of Innovation in the Commercial Aircraft Jet Engine Market," *Technological Forecasting and Social Change,* 1971. Also, some industrial consultants have reported that this model has been useful in helping clients with problems in this area.

have a significant effect on the rate of diffusion. To improve its performance as an analytical tool and a forecasting device, we extended this model in three major ways in this book. Our results are discussed in the following two sections.

13. Industrial Organization and the Diffusion Process

The first way that we extended the basic diffusion model was to add the market structure of the industry as an additional explanatory variable. In our study of the diffusion of numerically controlled machine tools, we tried to estimate the effects of industrial organization on an industry's rate of diffusion. In other words, does it appear, when other relevant variables are held constant, that numerically controlled machine tools spread more rapidly in more concentrated or less concentrated industries? There has been considerable speculation concerning the effects of industrial concentration on the rate of diffusion, but little or no agreement has been reached on *a priori* grounds. What little empirical evidence we have from previous studies seems to indicate that innovations tend to spread more rapidly in less concentrated industries, but the data have been too sparse to warrant any conclusive statement.[40]

Our results in this book seem to indicate that the rate of diffusion is definitely higher in less concentrated industries, at least in the case of numerically controlled machine tools. Both the number of firms in the industry and the inequality of firm sizes have a statistically significant effect on the rate of diffusion. Of course, these results only pertain to a single innovation, numerically controlled machine tools. But they are quite consistent with previous results pertaining to other innovations. The fact that innovations tend to spread more rapidly in less concentrated industries is particularly noteworthy, given the common impression that industries composed of a few giant firms tend to be technologically most progressive.

As in previous studies, our results indicate too that, in a given industry, the larger firms tend to be quicker than the smaller ones to begin using numerical control. As we have pointed out so frequently before, this does not mean that the larger firms are more progressive than the smaller ones. Even if the larger firms are no more progressive than the small ones, we might expect them to be quicker, on the average, to begin using a new technique.[41] In the chemical industry, this same sort of relationship between a firm's size and how rapidly it began using a new technique was found for the more costly innovations, but much less often for the others. It seems reasonable to expect that a relationship of this sort will occur less often among relatively inexpensive innovations.

40. Mansfield et al., *op. cit.,* and Mansfield, *op. cit.*

41. *Ibid.*

A firm's size may also have an effect on its intrafirm rate of diffusion of an innovation. Our results concerning numerical control indicate that, in the ten industries studied, a firm's intrafirm rate of diffusion is inversely related to its size. This is quite consistent with our earlier findings concerning the intrafirm rates of diffusion of the diesel locomotive in the railroad industry.[42] In the chemical industry, the intrafirm rate of diffusion is often almost instantaneous, since one process is often substituted entirely for another. In cases where this is not the case, the intrafirm rate of diffusion seems to be inversely related to a firm's share of the market for the relevant product. Thus, with regard to all the innovations studied (in the chemical and other industries) where intrafirm rates of diffusion vary among firms, there is a tendency for firms with large shares of the market to have relatively low intrafirm rates of diffusion.

14. R and D Expenditures, New Products, and the Diffusion Process

The second way that we extended our basic diffusion model was to include the effects of an industry's R and D expenditures (as a percent of sales) on its rate of diffusion of innovations. Although it seems reasonable to believe that research-intensive industries will be quicker to accept and utilize new technology than industries that do relatively little R and D, previous studies have made no attempt to test this proposition. In this work, we carried out such a test for one important manufacturing innovation where we could obtain the required data. Specifically, in our study of numerically controlled machine tools, we included an industry's R and D expenditures (as a percent of sales) as one of the determinants of its rate of diffusion. In accord with this hypothesis, our results indicate that this variable had a statistically significant effect.

These findings are interesting not only because they shed new light on a factor that has tended to be neglected in previous discussions of the diffusion process; in addition, they are interesting because they suggest that the role of R and D is broader than economists often presume. In their models, economists view R and D as basically an invention-producing activity. Although this certainly is part of what R and D does, it is by no means its only mission. In addition, R and D provides the firm with a window opening on various parts of its environment. For example, it allows the firm to communicate more readily with the scientific and technological community, and to understand and evaluate other firms' innovations more quickly and more accurately. Also, it sometimes is a device to recruit and train people who eventually will move on to general management. Given these functions of R and D, it is not surprising that more R and D-intensive industries tend to accept innovations more rapidly than others.

42. E. Mansfield, *Industrial Research and Technological Innovation, op. cit.*

The third extension of our basic diffusion model was to look at the determinants of the rate of diffusion of product innovations. Practically all of the results concerning the diffusion of innovations pertain to processes, not products.[43] We know very little about the factors determining how quickly a new product will be imitated by other firms. Yet this clearly is a very important part of the process of industrial competition. To throw some much-needed light on this kind of imitation (or entry) process, Simon studied the rate at which 29 new products in the chemical industry were imitated. The results indicated that the profitability of the innovation, the size of the investment required to produce the innovation, and the existence and duration of patents have statistically significant effects on the rate of imitation. Indeed, a simple econometric model based on these variables can explain about two-thirds of the observed variation in the rates of imitation.

These results seem to be the first quantitative estimates of the effects of various determinants of the rate of imitation of a new product. Among other things, they are of interest because they demonstrate the basic similarities between the imitation process for products and that for processes: In both cases, essentially the same variables—the profitability of the innovation, the size of the investment required, and the existence and duration of patents (as well as a trend factor)—seem to be the basic determinants of the rate of imitation. Also, it is interesting to note that the results are quite consistent with an earlier study by one of the authors of entry rates.[44]

15. International Technology Transfer

Technology spreads from country to country, as well as from firm to firm. According to some indicators, the relative position of the United States as a source of major technological innovations has been declining. For example, the 1975 annual report of the National Science Board contains a study indicating that the percentage of major innovations originating in the United States fell substantially from 1953 to 1967, and has remained relatively stable since 1968. Of course, there are a host of conceptual and statistical difficulties with measures of this sort, as the National Science Board recognizes. And to some extent, a dimi-

43. There have been a number of studies of the growth of sales of consumer products. This is not what we have in mind. What we mean is that there have been practically no studies of the rate at which firms begin producing a product that is essentially the same as a new product introduced by an innovator.

44. E. Mansfield, "Entry, Gibrat's Law, Innovation, and the Growth of Firms," *American Economic Review,* December 1962. In this paper, too, profitability and capital requirements are found to be significant variables in determining entry rates.

nution in our technological lead would have been expected, as countries like West Germany and Japan got back on their feet after the war. Nonetheless, to the extent that these indicators are valid, and their past trends continue, they may portend problems in international markets for some of our industries that currently contribute greatly to our trade balance.[45]

Our results provide a modest amount of information bearing on this topic. In particular, they show that a substantial percentage of the chemical innovations in Chapter 3 were based directly on foreign developments. This is an important fact, because some groups, under the misapprehension that the United States has a monopoly on advanced technology, have suggested that we try to interfere with the international transfer of technology in order to prevent this technology from seeping out to competitors in other countries. Such a policy would suffer from major disadvantages, since it would be difficult to accomplish and, even if it could be done, it would invite retaliation. As is evident from our results, there are mutual benefits from international specialization with regard to technology, and technology flows both out of and into the United States.[46]

Some observers have also argued that the government should support and stimulate R and D in industries where our technological lead is narrowing. In the short run, our nation's policies with regard to exchange rates, tariffs, and inflation are likely to have more effect on our trade balance than our policies with respect to science and technology. In the long run, the international competitiveness of particular American industries most certainly will depend on our policies with respect to science and technology, but one cannot conclude from this that government support for civilian technology should be focused on industries that are having increased difficulties in meeting foreign competition. Whether or not more R and D should be supported or encouraged in a particular industry depends on the extent of the social payoff there from additional R and D, not on whether or not our technological lead there seems to be shrinking. In other words, it depends on the social rate of return, as we pointed out at the beginning of this chapter.

45. See *Science Indicators, 1974* (Washington, D.C.: National Science Foundation, 1975); R. Gilpin, *Technology, Economic Growth, and International Competitiveness* (Washington, D.C.: Joint Economic Committee of Congress, 1975); and E. Mansfield, *Federal Research and Development Expenditures and the National Economy,* testimony before the House Committee on Science and Technology, April 28, 1976.

46. For some relevant discussion, see E. Mansfield, "International Technology Transfer: Problems, Costs, Policies," *American Economic Review* (May 1975); E. Mansfield, "Technology and Technological Change," in J. Dunning, ed., *Economic Analysis and the Multinational Enterprise* (London: George Allen and Unwin, 1974); and National Science Foundation, *The Effects of International Technology Transfers on U.S. Economy* (Washington, D.C.: Government Printing Office, 1974).

16. Limitations of the Studies

In conclusion, we should add the customary warning that the studies contained in this volume are based on only a sample of industries—chemicals, petroleum, drugs, electronics, machine tools, aircraft engines, aircraft, farm machinery, instruments, printing presses, tools and dies, farm equipment, electrical equipment, coal mining machinery, metals, construction, paper, apparel, and household cleaners. Moreover, within each of these industries, the data pertain generally to only a sample of firms. Because the basic data are limited in this way and because the models used here are rough in many respects, our results must be regarded as tentative. Much more work is required before we attain a satisfactory understanding of the production and diffusion of new technology in the entire range of American industry.

On the other hand, it is our belief that these studies have gone quite far beyond anything attempted heretofore in analyzing the relevant aspects of the process of technological change. Our findings are based on data obtained from literally hundreds of firms, many of whose engineers, scientists, and managers spent considerable amounts of time with us. We have obtained extremely detailed data concerning hundreds of R and D projects, as well as the decisions and activities involved in these projects. New types of data have been gathered, and new models have been proposed, to help answer a variety of questions, many of which were previously almost totally unexplored. The studies described here are part of a continuing study of the economics of technological change. We hope to extend and deepen them—and the other investigations we have carried out in this area—in the future.

INDEX